The IMA Volumes
in Mathematics
and its Applications

Volume 39

Series Editors
Avner Friedman   Willard Miller, Jr.

Institute for Mathematics and
its Applications
IMA

The **Institute for Mathematics and its Applications** was established by a grant from the National Science Foundation to the University of Minnesota in 1982. The IMA seeks to encourage the development and study of fresh mathematical concepts and questions of concern to the other sciences by bringing together mathematicians and scientists from diverse fields in an atmosphere that will stimulate discussion and collaboration.

The IMA Volumes are intended to involve the broader scientific community in this process.

Avner Friedman, Director
Willard Miller, Jr., Associate Director

* * * * * * * * * *

IMA ANNUAL PROGRAMS

1982–1983	**Statistical and Continuum Approaches to Phase Transition**
1983–1984	**Mathematical Models for the Economics of Decentralized Resource Allocation**
1984–1985	**Continuum Physics and Partial Differential Equations**
1985–1986	**Stochastic Differential Equations and Their Applications**
1986–1987	**Scientific Computation**
1987–1988	**Applied Combinatorics**
1988–1989	**Nonlinear Waves**
1989–1990	**Dynamical Systems and Their Applications**
1990–1991	**Phase Transitions and Free Boundaries**
1991–1992	**Applied Linear Algebra**
1992–1993	**Control Theory and its Applications**
1993–1994	**Emerging Applications of Probability**

IMA SUMMER PROGRAMS

1987	**Robotics**
1988	**Signal Processing**
1989	**Robustness, Diagnostics, Computing and Graphics in Statistics**
1990	**Radar and Sonar**
1990	**Time Series**
1991	**Semiconductors**
1992	**Environmental Studies: Mathematical, Computational, and Statistical Analysis**

* * * * * * * * * *

SPRINGER LECTURE NOTES FROM THE IMA:

The Mathematics and Physics of Disordered Media

Editors: Barry Hughes and Barry Ninham
(Lecture Notes in Math., Volume 1035, 1983)

Orienting Polymers

Editor: J.L. Ericksen
(Lecture Notes in Math., Volume 1063, 1984)

New Perspectives in Thermodynamics

Editor: James Serrin
(Springer-Verlag, 1986)

Models of Economic Dynamics

Editor: Hugo Sonnenschein
(Lecture Notes in Econ., Volume 264, 1986)

F. Alberto Grünbaum   Marvin Bernfeld
Richard E. Blahut
Editors

Radar and Sonar

Part II

With 52 Illustrations

Springer-Verlag
New York Berlin Heidelberg London Paris
Tokyo Hong Kong Barcelona Budapest

F. Alberto Grünbaum
Department of Mathematics
University of California
Berkeley, CA 94720
USA

Marvin Bernfeld
The Raytheon Company
430 Boston Post Road
Wayland, MA 01778
USA

Richard E. Blahut
IBM
Owego, NY 13827-1298
USA

Series Editors
Avner Friedman
Willard Miller, Jr.
Institute for Mathematics and its
    Applications
University of Minnesota
Minneapolis, MN 55455
USA

Mathematics Subject Classification: 78A45, 22E70, 43A80, 65799

Library of Congress Cataloging-in-Publication Data
(Revised for volume II)
Blahut, Richard E.
  Radar and sonar.
  (IMA volumes in mathematics and its applications;
v. 32, 39)
  Vol. 2 edited by Alberto Grünbaum, Marvin Bernfeld,
and Richard Blahut.
  Includes bibliographical references.
  1. Radar — Mathematical models.   2. Sonar — Mathemati-
cal models.   I. Miller, Willard.   II. Wilcox, Calvin H.
(Calvin Hayden).   III. Title.
TK6580.B5   1991   621.3848   90-27282
  ISBN 978-1-4684-7834-1   ISBN 978-1-4684-7832-7 (eBook)
  DOI 10.1007/978-1-4684-7832-7

Printed on acid-free paper.

ISBN 978-1-4684-7834-1
Production managed by Bill Imbornoni.
Camera-ready copy prepared by the IMA.

9 8 7 6 5 4 3 2 1
ISBN 978-1-4684-7834-1

The IMA Volumes
in Mathematics and its Applications

Current Volumes:

Forthcoming Volumes:

1989-1990: *Dynamical Systems and Their Applications*

    Partial Differential Equations with Minimal Smoothness and Applications

    Twist Mappings and Their Applications

    Dynamical Theories of Turbulence in Fluid Flows

    Nonlinear Phenomena in Atmospheric and Oceanic Sciences

    Chaotic Processes in the Geological Sciences

Summer Program 1990: *Time Series in Time Series Analysis*

    Time Series (2 volumes)

1990-1991: *Phase Transitions and Free Boundaries*

    On the Evolution of Phase Boundaries

    Shock Induced Transitions and Phase Structures

    Microstructure and Phase Transitions

    Statistical Thermodynamics and Differential Geometry
    of Microstructured Material

    Free Boundaries in Viscous Flows

Summer Program 1991: *Semiconductors*

    Semiconductors (2 volumes)

FOREWORD

This IMA Volume in Mathematics and its Applications

RADAR AND SONAR, PART II

is based on the proceedings of the second week of the IMA summer program "Radar and Sonar". Week two was devoted to the presentation of problems and solutions of problems in Radar and Sonar. We are grateful to Alberto Grünbaum, Marvin Bernfeld and Richard Blahut for organizing the program.

We also take this opportunity to thank those agencies whose financial support made the summer program possible: the Air Force Office of Scientific Research, the Army Research Office, the National Science Foundation, the National Security Agency and the Office of Naval Research.

Avner Friedman

Willard Miller, Jr.

PREFACE

This volume contains a representative discussion of mathematical problems that arise in radar and sonar and is based on the lectures that were given during the second week of the IMA summer program RADAR AND SONAR, June 18–June 29, 1990. (The first week was devoted to three sets of tutorial lectures and the lecture notes from that part of the program appear in an earlier IMA volume.) The second week was run as a workshop of contributed papers without formal review. The speakers were selected to cover a broad range of problems in this area.

The summer program was organized to stimulate a dialogue between engineers and applied mathematicians. The design of waveforms for radar and sonar and the development of algorithms for the processing of these waveforms lead to many interesting and difficult problems of applied mathematics. It is timely to separate these problems from the engineering tasks of radar and sonar so as to form a minitopic of applied mathematics. The range of such problems contained herein probably cannot be found in any other volume. There are applications of group theory, modern topics of signal processing, inverse problems, array processing and beamforming, estimation-theoretic imaging, and phase tracking. We believe the program was a great success. As these and related topics develop further a future sequel to this program will be a success as well.

We take the opportunity to thank IMA director Avner Friedman, associate director Willard Miller, Jr. and the exemplary staff of the IMA, and to also thank Ms. Patricia Brick, Mr. Stephen Skogerboe, and Ms. Kaye Smith for the preparation of the manuscripts.

F. Alberto Grünbaum
Marvin Bernfeld
Richard E. Blahut

CONTENTS

BEAMFORMING WITH DOMINANT MODE REJECTION

DOUGLAS A. ABRAHAM* AND NORMAN L. OWSLEY*

Abstract. High resolution adaptive beamforming of a spatial array of sensors has traditionally been formulated in terms of a minimum variance, distortionless response optimality criterion. This paper develops *enhanced* minimum variance distortionless response beamformers from the eigenstructure of the *array cross-spectral density matrix*.

Introduction. High resolution adaptive beamforming of a spatial array of sensors has traditionally been formulated in terms of a minimum-variance, distortionless response (MVDR) optimality criterion. This approach results in a constrained spatial Wiener filter structure which requires the estimation and inversion of either the array space-time correlation matrix for time-domain implementation or the array cross-spectral density matrix (CSDM) for a frequency domain implementation [1]. In the situation where the array acoustic environment consists of a limited number of dominant modes the full matrix inversion procedure is both numerically and statistically inadvisable. In this case, the technique of point-source interference null steering can be generalized to the nulling of dominant modes without prior knowledge of the mode structure. In effect, a reduced rank version of the spatial Wiener filter is produced which is realized as a generalized mode-nulling beamformer.

This paper develops the dominant mode rejection or enhanced minimum-variance distortionless response [2] (EMVDR) beamformer from the eigenstructure of the array CSDM. The EMVDR concept is similar to work done by Kirsteins and Tufts [3]. The MVDR criterion of adaptive beamforming requires the estimation and inversion of the array CSDM. The EMVDR beamformer suggests the approximation of the CSDM by the enhanced dominant, i.e. interfering, plane-wave source subspace. This only requires the estimation of the eigenstructure associated with these dominant modes. The dominant eigenstructure form of the approximated CSDM allows simple computation of the beamforming filter vector avoiding the inversion process required of the MVDR beamformer.

The performance of the EMVDR beamformer is compared to that of the MVDR beamformer in terms of array gain improvement for the case of a single plane wave source, a single plane wave or point interference and uncorrelated background noise. The EMVDR beamformer is then implemented through three different eigenstructure estimators. The beam responses of a conventional beamformer (CBF) and an MVDR beamformer are compared to those of the various EMVDR implementations.

---

*Naval Underwater Systems Center, New London, Connecticut 06320.

Development of EMVDR Beamformer. The array CSDM is the expected value of the outer product,

$$R = E[x(k,f)x^H(k,f)], \tag{1}$$

where $x(k,f)$ is the frequency domain data vector for time sample k and frequency f, and the superscript "H" denotes the complex conjugate and transpose operation.

When the array acoustic environment includes several plane-wave sources, angularly extended interferences and background spatially uncorrelated noise, the CSDM for an array of N sensors has the form,

$$R = DED^H + Q + \sigma^2 I_N, \tag{2}$$

where $D = [d_1 d_2 \cdots d_P]$ is a matrix whose columns are the array steering vectors associated with the P plane-wave sources, E is a matrix of the source covariances and is diagonal for uncorrelated sources, Q is a matrix of the covariances of the angularly extended sources, σ^2 is the uncorrelated noise variance at a single sensor and I_N is the N-by-N identity matrix.

The eigendecomposition of the CSDM is written as,

$$R = \sum_{i=1}^{N} \lambda_i m_i m_i^H, \tag{3}$$

where λ_i are the eigenvalues and m_i are the eigenvectors of R.

If $\lambda_1 \geq \lambda_2 \geq \cdots \geq \lambda_N$, then let the first D eigenvalues be predominantly due to strong plane waves, and the rest due to weaker plane waves, angularly extended and spatially uncorrelated noises. The CSDM may then be approximated by the dominant modes corresponding to the D strong plane waves,

$$\widehat{R}(e) = e \sum_{i=1}^{D} \psi_i m_i m_i^H + \sigma^2 I_N, \tag{4}$$

where $\psi_i = \lambda_i - \sigma^2$ and e is a scalar enhancement factor.

The beamforming vector is formed by placing the approximated CSDM into the Weiner filter equation,

$$\hat{w}(\theta, e) = \frac{\widehat{R}^{-1}(e)d(\theta)}{d^H(\theta)\widehat{R}^{-1}(e)d(\theta)}. \tag{5}$$

Substituting the dominant eigenstructure representation of the approximated CSDM from Equation 4 into the beamforming vector in Equation 5 results in the revealing form,

$$\hat{w}(\theta, e) = \frac{d(\theta) - \sum_{i=1}^{D} \beta_i m_i m_i^H d(\theta)}{N - \sum_{i=1}^{D} \beta_i |m_i^H d(\theta)|^2}, \tag{6}$$

where $\beta_i = \frac{1}{1 + \frac{\sigma^2}{e\psi_i}}$. The beamforming vector is seen to be a conventional array steering vector less the projection of that look direction onto the eigenvectors associated with the dominant modes with a normalization term in the denominator. Thus the EMVDR beamformer is a generalized mode-nulling beamformer where no prior knowledge of the mode structure is required.

Array Gain Improvement. For the case of a single plane-wave source and single plane-wave or point interference amidst spatially uncorrelated background noise the CSDM has the form,

$$R = s_1 d_1 d_1^H + s_2 d_2 d_2^H + \sigma^2 I_N, \tag{7}$$

where s_i and d_i are respectively the signal and interference powers or variances and array steering vectors.

The array gain (AG) is defined as the ratio of the signal-to-noise ratio (SNR) at the output of a beamformer to the SNR at a single sensor. The array gain improvement (AGI) is defined [2,4] as the ratio of the array gain for a particular adaptive beamformer to the array gain of a conventional beamformer,

$$AGI = \frac{AG_{ABF}}{AG_{CBF}}. \tag{8}$$

For the case of a single source and interference described above, the array gain improvement may be found analytically for the EMVDR beamformer using only one dominant mode to be,

$$AGI_{EMVDR} = \frac{[(u - L) + (u - 1)\gamma L(2 + \gamma)]^2 (1 + rL)}{\{(1 + rL)[u(1 + \gamma L) + (u - 1)\gamma L(1 + \gamma)]^2 \atop -L(r + 1)(1 + \gamma)[(2u - 1)(1 + 2\gamma L + \gamma^2 L) - \gamma(1 + \gamma L)]\}} \tag{9}$$

where,

$$\gamma = \frac{2p}{(1 - p) + \sqrt{(1 - p)^2 + 4pL}}$$

$$p = \text{signal-to-interference ratio} = \frac{s_1}{s_2}$$

$$L = \frac{|d_2^H d_1|^2}{N^2}$$

r = CBF beam output interference to

background noise ratio $= \dfrac{N s_2}{\sigma^2}$

$$u = 1 + \frac{\gamma}{erp(1 + \gamma)}$$

e = dominant mode enhancement factor.

Note that when the enhancement is increased to infinity, $u = 1$ and the EMVDR beamformer produces a MUSIC-like [5] angular response spectrum. In the same context the array gain improvement for the element space MVDR beamformer [4] is,

$$AGI_{MVDR} = 1 + \frac{r^2 L(1 - L)}{1 + r}. \tag{10}$$

The array gain improvement for the EMVDR and MVDR beamformers is compared in Figures 1 – 3 for three different signal-to-interference (SIR) ratios (respectively $-5, -10,$ and $-20dB$). The array gain improvement is plotted against

r, the conventional beamformer beam output interference-to-noise ratio. On each graph, the solid lines trace the MVDR array gain improvement and the dotted lines trace the conventional beamformer array gain for several values of L ($-3, -10, -20,$ and $-30dB$), the normalized sidelobe level of the interference in a beam steered to the signal. The dashed lines in Figures 1 and 2 and the x's in Figure 3 represent the EMVDR array gain improvement with unity enhancement. The conventional beamformer array gain is shown for comparison and so the overall array gain of an adaptive beamformer may be computed from the graph. Some interesting results are drawn from these figures as follows.

- With only one dominant mode, the EMVDR beamformer can equal the performance of the MVDR beamformer for the important case of low signal-to-interference ratios as seen in Figure 3 and for low r in Figure 2.

- At higher signal-to-interference ratios the first dominant mode begins to contain more information about the signal. This causes the degradation seen in the EMVDR array gain improvement in Figures 1 and 2.

- At high values of r the signal is also substantially above the background noise, thus a second dominant mode should be used. The effect of using only one dominant mode is seen in Figures 1 and 2 at high r where the EMVDR array gain improvement levels off. The limiting value of the array gain improvement for high r is found by letting r go to infinity in Equation 9 and results in,

(11)
$$\text{Cutoff} = \frac{1}{\gamma^2}.$$

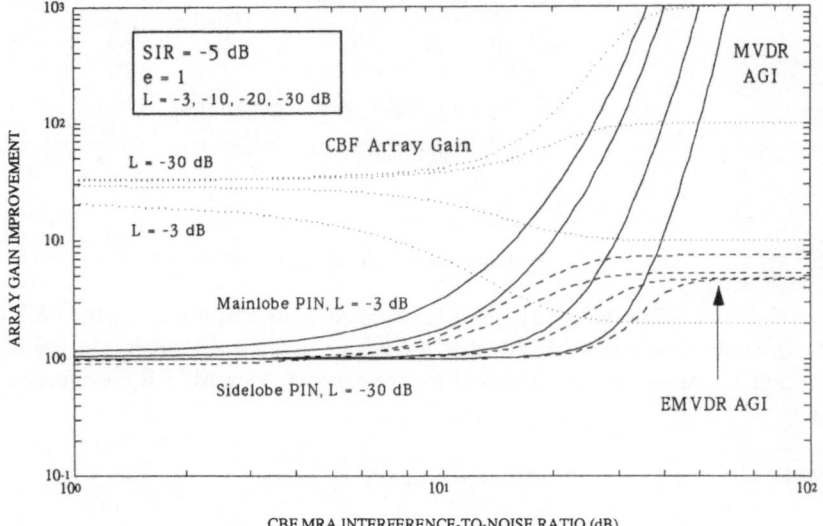

Figure 1 MVDR and EMVDR AGI for SIR $= -5dB$.

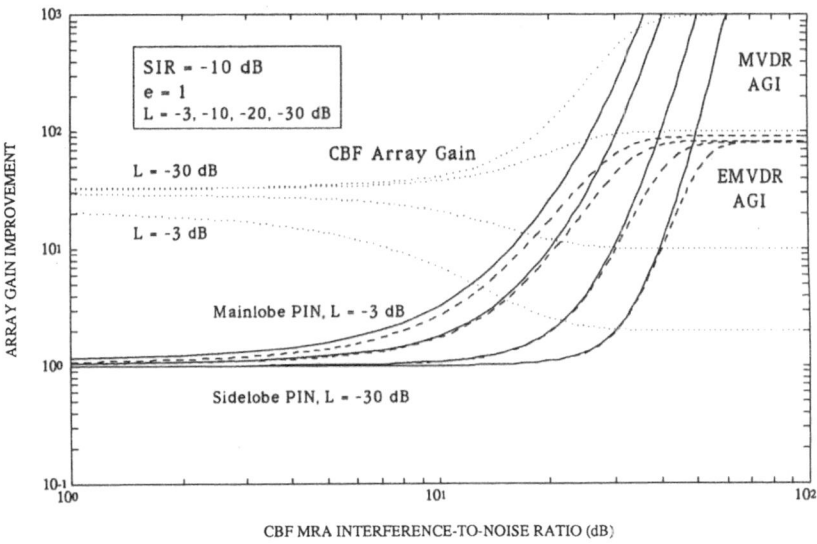

Figure 2 MVDR and EMVDR AGI for SIR $= -10db$.

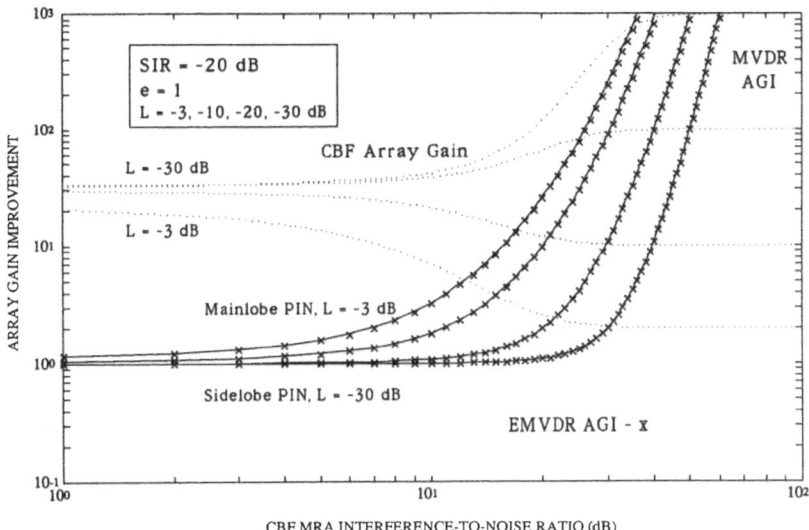

Figure 3 EMVDR and MVDR AGI for SIR $= -20db$.

Implementation. One of the objectives in using the EMVDR beamformer is to avoid the numerical intensity required of the matrix inversion in the MVDR beamformer. This section uses three different eigenstructure estimation schemes to implement the EMVDR beamformer and compares their angular beam responses to those of a conventional beamformer and MVDR beamformer with simulated data. The beam responses are displayed both in a waterfall plot over time and overlaid to provide an average beam response for the stationary scenarios.

The baseline eigenstructure estimator is a brute force International Mathematical & Statistical Libraries, Inc. (IMSL) subroutine that computes all of the eigenvalues and vectors of the exponentially time-averaged CSDM. This method is numerically more intensive than the latter two, a method published by Yang and Kaveh [6] and a gradient adaptive data orthogonalization technique (ADO) suggested by Owsley [7], which are numerically "quicker" than the IMSL subroutine. The fast algorithms exploit the structure of the CSDM and only estimate the D largest eigenvalues and vectors. The eigenvalue estimates for the quick methods are formed from the eigenvector estimates and an exponentially averaged CSDM.

Two cases are explored in this section to illustrate some of the properties of the EMVDR beamformer. The first case (Figures 4a-4h) is three stationary (in bearing and power) plane waves sources amidst background uncorrelated noise. The source of concern is at zero degrees and $-10dB$. The first interference is placed in the mainlobe of a conventional beamformer beam pointed at the source, at 10 degrees and 0 dB. The second interference is placed in a sidelobe, -30 degrees and $-3dB$. Figures 4a and 4b respectively have the conventional beamformer and MVDR beam responses. The IMSL eigenstructure EMVDR implementation (IMSL) is shown in Figures 4c-4f with respectively one, two and three dominant modes with unity enhancement and then with three dominant modes and an enhancement factor of ten. The Yang and Kaveh algorithm (Y& K) and the ADO eigenstructure implementations are shown in Figures 4g and 4h with three dominant modes and an enhancement factor of ten.

The second scenario consists of one source that is stationary in bearing and one that is moving. The stationary source is at -30 degrees at $0db$ and the moving source is at $-3dB$ and goes from -20 degrees to 0 degrees in 20 time samples, a rate of one degree per time sample. The power of both sources is constant. Figures 5a, 5b, and 5c are respectively the conventional beamformer, MVDR, and EMVDR beamformers where the EMVDR has the IMSL implementation with two dominant modes and an enhancement of one.

In each case the background noise has a sensor level variance of $-3dB$, there are 16 sensors in the array, and the conventional beamformer has $30dB$ sidelobe Chebyshev shading of the sensor data. A few interesting points are worth elaborating on as noted below:

- The low-level source of concern at 0 degrees and $-10dB$ is only clearly seen in 4f, 4g, and 4h where the EMVDR beamformer had three dominant modes and an enhancement factor of ten. The source is noticeable in the overlaid plots of all of the adaptive beamformers 4b-4h. The conventional beamformer fails to distinguish the low-level source from the mainlobe interference.

- The conventional beamforming of the source at -30 degrees with nulling of the source at 10 degrees is apparent in Figure 4c where only one dominant mode is used. This is compared to the higher resolution obtained in Figures 4d-4f where two or three dominant modes are used. The conventional beamformer with null steering of the dominant modes is typical of the EMVDR beamformer and obviously seen here for a strong source, but also occurs for the important case of low-level sources.

- As seen by comparing 4f to 4e, increasing the enhancement factor improves the resolution of the dominant plane waves in the source subspace.

- The Yang and Kaveh eigenstructure estimation algorithm of 4g provided high resolution of the strong interferences and also clearly shows the weaker source at zero degrees. The ADO algorithm in 4h showed a near MVDR like performance with a slightly greater resolution (due to the enhancement factor of ten).

- The second case illustrates how the movement of the $-3dB$ source is captured by the dominant eigenstructure of the CSDM. The EMVDR beamformer in Figure 5c is almost identical to the MVDR beamformer of Figure 5b. The time constant on the exponential filter used to estimate the CSDM was $\alpha = 0.9$ for 5b and $\alpha = 0.7$ for 5c, thus the EMVDR beamformer required a shorter time constant, i.e. faster adaptation, than the MVDR beamformer in order to achieve equivalent performance.

8

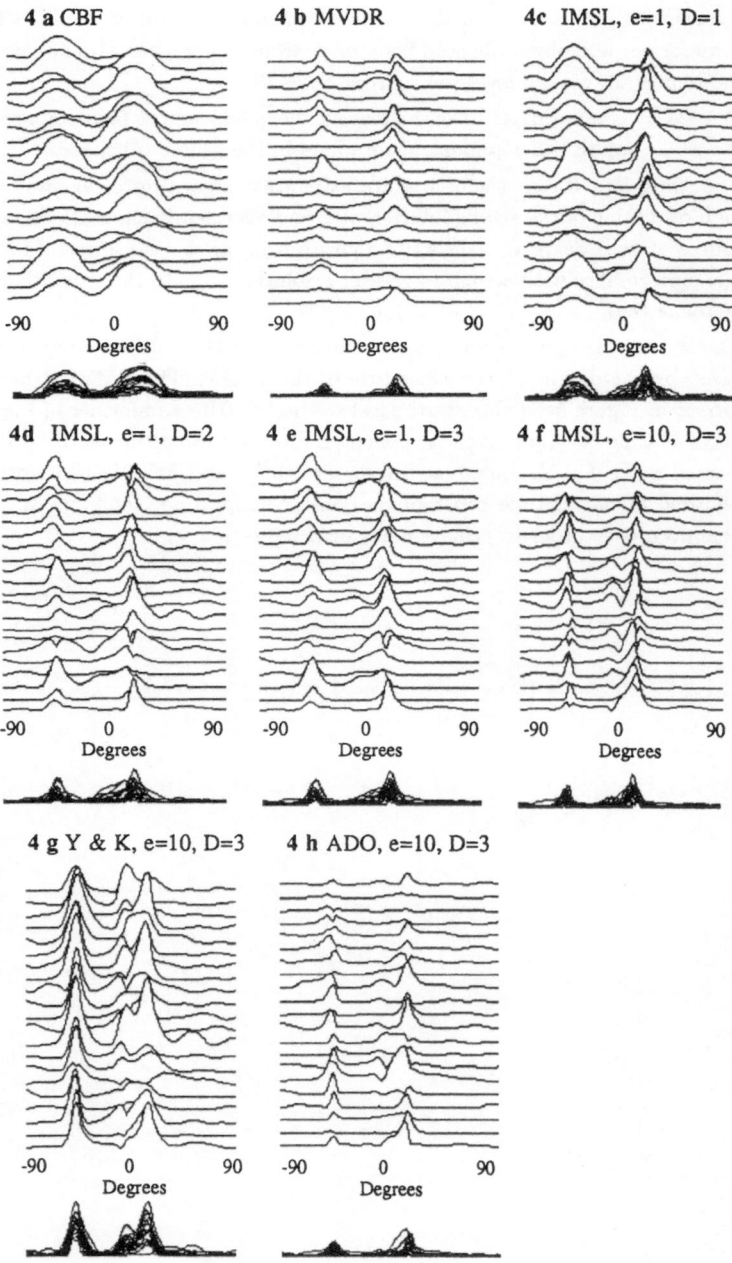

Figure 4a–4h Beam response plots for CBF 4a, MVDR 4b, and EMVDR 4c–4h beamformers. Plane waves at 0 degrees at $-10dB$, 10 degrees at $0dB$, -30 degrees at $-3dB$, background noise at $-3dB$, 16 sensor array.

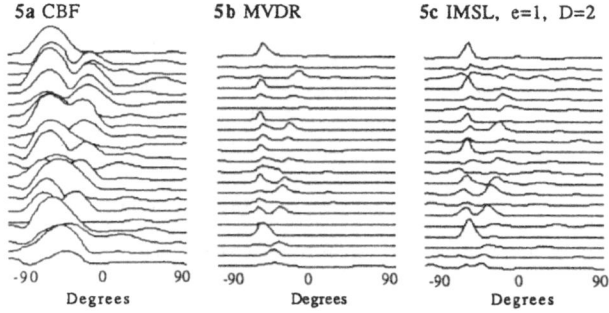

5a CBF   5b MVDR   5c IMSL, e=1, D=2

Figure 5a–5c Beam response plots for CBF 5a, MVDR 5b, and EMVDR 5c beamformers. One stationary plane wave at -30 degrees at $0dB$ and one moving at one degree per time sample from -20 to 0 degrees at $-3dB$.

Conclusions. The EMVDR beamformer has been shown to be a generalized mode-nulling beamformer that is based on an enhanced approximation to the array CSDM by its D dominant modes. The inversion of the CSDM required by the MVDR beamforming algorithm is avoided by the use of the eigenvalues and orthonormal eigenvectors of the array CSDM. The trade-off therefore exists between the estimation and inversion of the N-by-N CSDM and the estimation of the D dominant eigenvalues and vectors.

The array gain improvement of the EMVDR beamformer using one dominant mode has been shown to equal that of the MVDR beamformer for a case with two sources when one source is weak compared to the other. This may be extrapolated to mean that by using one dominant mode for each strong plane-wave interference, the EMVDR beamformer will achieve the same average array gain improvement as the MVDR beamformer.

The numerical savings and the performance of the EMVDR beamformer are dependent on the choice of eigenstructure estimator. The best performance was obtained with a brute force estimator that determines all of the eigenvalues and vectors but is numerically intensive. Quick methods that exploit the special form of the CSDM and approximate the eigenstructure are desirable for their numerical simplicity and have been shown to provide adequate – though not equivalent – performance. The critical choice of the eigenstructure estimator thus lies between these two opposing desires; fewer numerical operations or better performance.

Other interesting and desirable characteristics of the EMVDR beamformer include it's conventional beamformer property with dominant mode nulling. This provides optimal beamforming of low–level sources once the dominant interferers have been nulled. The varying enhancement factor also allows the beamformer to provide very high resolution spectral responses.

REFERENCES

[1] OWSLEY, N., *Array Signal Processing*, Ed. S. Haykin, Prentice-Hall (1985).
[2] OWSLEY, N., *Enhanced Minimum Variance Beamforming*, in *Underwater Acoustic Data Processing*, Ed. Y.T. Chan, Kluwer Academic Publishers, 1989 or NUSC Tech. Rpt. 8305, 18 Nov. 1988.
[3] KIRSTEINS, I. AND TUFTS, D., *Rapidly Adaptive Nulling of Interference*, in Springer-Verlag Series on Detection and Estimation (1989).
[4] OWSLEY, N. AND ABRAHAM, D., *Preprocessing for High Resolution Beamforming*, Proc. 23rd Asilomar Conference, Nov. 1989.
[5] SCHMIDT, R., *Multiple Emitter Location and Signal Parameter Estimation*, Proc. RADC Spectral Est. Workshop, Rome Air Development Center, Rome, NY, 1979.
[6] YANG, J. AND KAVEH, M., *Adaptive Eigensubspace Algorithms for Direction or Frequency Estimation and Tracking*, IEEE Trans. on Acoustics, Speech, and Signal Processing, Vol. 36, No. 2, Feb. 1988.
[7] OWSLEY, N., *Adaptive Data Orthogonalization*, Proc. IEEE ICASSP, Albuquerque, NM (1978).

FINE STRUCTURE OF THE CLASSICAL GABOR APPROXIMATION*

L. AUSLANDER[†],
M. AN, M. CONNER, I. GERTNER AND R. TOLIMIERI[‡]

Abstract. This paper presents some numerical experiments that help to understand the errors produced by replacing the exact Gabor expansion by a finite Gabor expansion, and so begin to expose some of the subtleties of Gabor approximations.

1. Introduction. Dennis Gabor in [7] introduced expansions of functions $f \in L^2(\mathbf{R})$ of the form

$$(1) \qquad f = \sum_m \sum_n a_{mn} e^{-\pi(t-n)^2} e^{2\pi i m t}$$

for the purpose of describing the energy distribution of the signal f over time and frequency. In this paper, after a brief review of the theory of Gabor expansions, we will present some numerical experiments that begin to expose some of the subtlety of the Gabor expansions. In particular, for digital computations we require an understanding of the errors produced by replacing the infinite sums in (1) by finite Gabor expansions.

We performed numerical experiments on the Hermite functions H_n, $n = 0, 1, \cdots, 12$. The numerical experiments were performed by sampling H_n at 1024 points and computing 1024 Gabor coefficients parameterized by $0 \leq m < 32$ and $0 \leq n \leq 32$. These *Gabor coefficients* synthesize, up to machine error, the sampled H_n. The cost of a digital Gabor computation depends on the number of these coefficients required to synthesize a sufficiently good approximation. The main result is that this cost is highly dependent on the eigenvalue space of the Hermite function relative to the Fourier transformation. For instance using standard L^2 error measurements we obtain the following results for H_4.

Using n Gabor coefficients	% error
$n = 15$	.261
$n = 27$	.027
$n = 51$	.0032
$n = 101$	.00026

This shows that by doubling the number of coefficients we reduce the error by a factor of 10.

* This research was supported by the advanced Research Projects Agency of the Department of Defense and was monitored by the Air Force Office of Scientific Research under contract number F49620-89-00020. The United States Government is authorized to reproduce and distribute reprints for governmental purposes notwithstanding any copyright hereon.

† DARPA, Washington D.C.

‡ Center for Large Scale Computation/CUNY, NY

In general, our results may be summarized as follows: By doubling the number of terms in the Gabor expansion we achieve the following improvement in the approximation.

$n \equiv 0 \bmod 4$ 10 fold improvement
$n \equiv 1 \bmod 4$ about 3/2 fold improvement
$n \equiv 2 \bmod 4$ 3 fold improvement
$n \equiv 3 \bmod 4$ 5 fold improvement.

We chose this numerical experiment guided by the theoretical results of the next section. The numerical experiments give an extraordinary demonstration of the implications of the theory for the special case we have studied. The relationship between Gabor expansions and Hermite expansions is of special interests in light of recent works by J-B. Martens [10].

The Zak transform [14] provides an important tool for manipulating Gabor expansions. In [3,4], a discrete version of the Zak transform was used to design several algorithms for carrying out Gabor computations with greater efficiency, accuracy, and stability than had previously been achieved by methods relying on biorthogonal expansions.

That Gabor expansions are related to nilpotent group theory has long been recognized. In [13], Weil established the significance of a transform similar to the Zak transform, called the Weil map, in Heisenberg group theory. Subsequent work [2,11,12] developed Gabor expansion theory and the related theory of the radar ambiguity function in the language of the Heisenberg group. The key impact of the Zak transform is that it decomposes the inherant nilpotency of Gabor expansions into two distinct abelian stages.

For any $g \in L^2(\mathbf{R})$, general Gabor expansions of the form

$$(2) \qquad f = \sum_m \sum_n a_{mn} g(t - n) e^{2\pi i m t}$$

were introduced in [6]. The Zak transform of g:

$$(3) \qquad Z(g)(x, y) = \sum_r g(y + r) e^{2\pi i r x}$$

can be used to establish conditions for the existence and uniqueness of Gabor expansions for any $f \in L^2(\mathbf{R})$. Many of the following results can be found in [8]. More recent results can be found in [5].

$A1$ If $Z(g)$ vanishes only on a set of measure 0 then the set of functions

$$(4) \qquad \{g_{mn} : m, n \in \mathbf{Z}\}$$

has dense span in $L^2(\mathbf{R})$ with $g_{mn}(t) = g(t - n) e^{2\pi i m t}$, $m, n \in \mathbf{Z}$. We will assume property $A1$ throughout the discussion. Because, as is well-known [1] the Gaussian, $g(t) = e^{-\pi i t^2}$, satisfies property $A1$, classical Gabor expansions always exist. The question of uniqueness is more complicated. Any $g_{m_0, n_0}, m_0, n_0 \in \mathbf{Z}$, can be removed from (4), without effecting the L^2-closure of the linear span. It follows that g_{m_0, n_0} has at least two Gabor expansions in $L^2(\mathbf{R})$.

We do know that any finite subset of (4) is linearly independent and that any $f \in L^2(\mathbf{R})$ admits at most one expansion (2) in $L^2(\mathbf{R})$ satisfying

$A2$ $\sum_m \sum_n | a_{mn} |^2 < \infty.$

However, not every $f \in L^2(\mathbf{R})$ admits a Gabor expansion satisfying property $A2$. Gabor expansions exist in $L^2(\mathbf{R})$ whose coefficients decay sufficiently slowly that $A2$ is violated. Consequently, significant information may be lost if, as is necessary for digital computation, we replace the infinite Gabor expansion by a truncated finite Gabor expansion. The main goal of the work is to provide tools for classifying this decay rate and to suggest modifications which enhance the efficiency of finite Gabor expansions to represent signal information.

Our method for constructing Gabor expansions is based on the observation [1,6] that if f admits a Gabor expansion satisfying $A2$, then the Gabor coefficients are given by Fourier coefficients of the doubly periodic function

$$(5) \qquad\qquad Z(f)/Z(g).$$

For arbitrary $f \in L^2(\mathbf{R})$, $Z(f)/Z(g)$ is not generally square-integrable over the unit square. However, if g is the Gaussian and $Z(f)$ is bounded then $Z(f)/Z(g)$ is integrable over this unit square and Fourier coefficients can be defined. These coefficients define a formal Gabor expansion for f. In contrast to standard Fourier analysis of signals, knowledge of the essential support of f and its Fourier transform does not solve the problem of determining a good finite Gabor expansion for f. The uncertainty principal plays a more critical role in simultaneous time-frequency expansions as compared to standard Fourier analysis representations. The emphasis in this work is on the obstruction coming from the Zero theorem [1,14].

2. Zak Transform Fundamental Formulas. The results of the preceding section are based on several well-known formulas involving Zak transforms. We shall briefly outline these formulas in this section. The Zak transform of a signal $f \in L^2(\mathbf{R})$,

$$(6) \qquad\qquad Z(f)(x,y) = \sum_r f(y+r)e^{2\pi i r x}$$

satisfies the functional equations

$$(7) \qquad\qquad Z(f)(x+1,y) \quad = Z(f)(x,y)$$
$$(8) \qquad\qquad Z(f)(x,y+1) \quad = e^{-2\pi i x} Z(f)(x,y).$$

It follows that $Z(f)$ is completely determined by its values on the unit square. If $H(1)$ denotes the Hilbert space of functions F on $\mathbf{R}^2$ satisfying (2) and (3) with inner product

$$(9) \qquad\qquad \|F\|_2^2 = \int_0^1 \int_0^1 | F(x,y) |^2 \, dx dy$$

then the mapping

$$(10) \qquad\qquad Z : L^2(\mathbf{R}) \longrightarrow H(1)$$

is an isometry of $L^2(\mathbf{R})$ onto $H(1)$.

Let $g \in L^2(\mathbf{R})$. The function

$$(11) \qquad\qquad g_{mn}(t) = g(t-n)e^{2\pi i m t}, m,n \in \mathbf{Z},$$

is a simultaneous time-frequency shift of g and as such, results from nilpotent action [1,11]. The Zak transform separates this action by the fundamental formula

$$(12) \qquad Z(g_{mn})(x,y) = Z(g)(x,y)e^{2\pi i(nx+my)}, m,n \in \mathbf{Z}.$$

We will assume throughout that $g \in L^2(\mathbf{R})$ and $Z(g)$ is bounded.

For any $f \in L^2(\mathbf{R})$, by (2) (3),

$$(13) \qquad Z(f)(x,y)Z^*(g)(x,y)$$

is a periodic function in both variables of period one. Set $I = [0,1)$ then

$$(14) \qquad \begin{aligned} \int_I \int_I &\mid Z(f)(x,y)Z^*(g)(x,y) \mid^2 dxdy \\ &\leq \sup_{I^2} \mid Z^*(g)(x,y) \mid^2 \cdot \|Z(f)\|_2^2 < \infty. \end{aligned}$$

Because Z is an isometry and

$$(15) \qquad \begin{aligned} < f, g_{mn} > &= < Z(f), Z(g_{mn}) > \\ &= \int_I \int_I Z(f)(x,y)Z^*(g)(x,y)e^{-2\pi i(nx+my)}dxdy \end{aligned}$$

we have the expansion of (8) as a Fourier series in $L^2(I^2)$

$$(16) \qquad Z(f)(x,y)Z^*(g)(x,y) = \sum_m \sum_n < f, g_{mn} > e^{2\pi i(nx+my)}$$

This is the main result needed to prove the comments about the dense span in $L^2(\mathbf{R})$ of the system $\{g_{mn}\}$ as well as the linear independence of any finite subset of this system.

Suppose

$$(17) \qquad f = \sum_m \sum_n a_{mn} g_{mn}$$

is a Gabor expansion in $L^2(\mathbf{R})$. Formally applying the Zak transform to both sides of (18) we have

$$(18) \qquad Z(f) = P\, Z(g)$$

where P is the trignometric series

$$(19) \qquad P = \sum_m \sum_n a_{mn} e^{2\pi i(nx+my)}.$$

P in this formal setting can be viewed as the Fourier series of the periodic function

$$(20) \qquad Z(f)/Z(g).$$

In general, the formal procedure can not be carried out. However, we do have the following result.

THEOREM 1. *If $f \in L^2(\mathbf{R})$ and*

$$(21) \qquad Z(f)/Z(g) \in L^2(I^2),$$

then there exists a unique Gabor expansion

(22) $$f = \sum_m \sum_n a_{mn} g_{mn},$$

satisfying property $A2$. The Gabor coefficients of (23) are given by the Fourier coefficients of $Z(f)/Z(g)$.

Conversely, if f admits a Gabor expansion satisfying $A2$ then $Z(f)/Z(g) \in L^2(I^2)$. These results can be found in [5]. Janssen [9] provides an account of uniqueness within the theory of distributions.

The obstruction to a simple Gabor theory comes from the Zero theorem [1].

Zero Theorem if $Z(g)$ is continuous then $Z(g)$ vanishes at some point in I^2.

For continuous $Z(g)$,we cannot assume $Z(f)/Z(g) \in L^2(I^2)$. This is independent of the smoothness or decay rate of f and $\hat{f}$ but depends ultimately on the zero set of $Z(f)$ cancelling the zero set of $Z(g)$. In general, when g is the Gaussian, we have that $Z(f)/Z(g)$ is integrable over the unit square and has a formal Fourier expansion

(23) $$\sum_m \sum_n a_{mn} e^{2\pi i(nx+my)}$$

The decay rate of the Fourier coefficients of (24) determines the degree of aliasing errors resulting from finite truncation of the corresponding Gabor expansion. A mixed time-frequency Nyquist sampling theorem is under investigation. In the next section, results introduced in [1] and [12], will be used to derive certain conditions which measure this aliasing.

3. Gaussian Case. Set $g(t) = e^{-\pi t^2}$, throughout this section. $Z(g)$ has a unique zero at $(\frac{1}{2}, \frac{1}{2})$ in I^2 and it is an analytic zero

(24) $$\frac{Z(g)(x,y)}{\left[(x - \frac{1}{2}) + i(y - \frac{1}{2})\right]} \in C^\infty(I^2).$$

The main result [1] follows.

THEOREM 2. *If $Z(f)$ is smooth*

- $\hat{f} = f$ implies $Z(f)/Z(g) \in C^1(T^2)$,

- $\hat{f} = if$ implies $Z(f)/Z(g) \in C^0(T^2)$,

- $\hat{f} = -f$ implies $Z(f)/Z(g) \in L^2(T^2)$,

- $\hat{f} = -if$ implies $Z(f)/Z(g) \in L^1(T^2)$,

where $T^2 = \mathbf{R}/\mathbf{Z}^2$. Extensions of this result can be found in [12].

Because $Z(f)/Z(g)$ is smoothest when $\hat{f} = f$, we would expect the decay rate of Gabor coefficients to be greatest in this case, with slower decay rate as we go down the list. A fixed number of (properly) choosen Gabor coefficients should contain the least information when $\hat{f} = -if$, with increasing information as we go up the list.

This Gabor analysis was carried out for the Hermite functions

$$(25) \qquad H_n(y) = c(n)(-1)^n e^{2\pi y^2} \frac{d^n(e^{-2\pi y^2})}{dy^n} e^{-\pi y^2},$$

where $c(n) > 0$ is taken such that $\|H_n\|^2 = 1, n \geq 0$. The Hermite functions are especially suited to test our results because

$$(26) \qquad \widehat{H}_n = (-i)^n H_n, \quad n \geq 0.$$

A recent work by Martens [10] suggests that Hermite expansions have several advantages over classically used Gabor expansions for processing visual data. This work is a first step in providing good Gabor or modified Gabor expansions, in a sense to be described below, for the Hermite functions, and as such, should give insights into the respective role played by Gabor and Hermite expansions.

For technical programming reasons the Hermite functions were offset and centered at $t = 16$. The actual reference function was

$$(27) \qquad g(t) = e^{-\pi(t-16)^2}$$

The reference function g and the 'Hermite' functions $H_n(t - 16)$ were set to zero outside the interval [0,32). The sample points were taken as

$$(28) \qquad \frac{2n+1}{64} + r, \quad 0 \leq n < 32, \quad 0 \leq r < 32.$$

The offset of $\frac{1}{64}$ was introduced to avoid the vanishing of the Zak transform of g at $(\frac{1}{2}, \frac{1}{2})$. For each $f = H_n(t - 16)$, $0 \leq n \leq 12$, we use the samples of f at the points (29) to compute samples of the Zak transform $Z(f)$ at the points

$$(29) \qquad \left(\frac{a}{32}, \frac{2b+1}{64} \right), \quad 0 \leq a < 32, \quad 0 \leq b < 32.$$

Because we have set f equal to zero outside the interval [0,32), these Zak transform samples can be computed by a 32-point FFT using the formula

$$(30) \qquad Z(f)\left(\frac{a}{32}, \frac{2b+1}{64} \right) = \sum_{r=0}^{31} f\left(\frac{2b+1}{64} + r \right) e^{2\pi i \frac{ra}{32}}, \quad 0 \leq a, b < 32.$$

Using the periodicity of these samples, we can write

$$(31) \qquad \frac{Z(f)}{Z(g)}\left(\frac{a}{32}, \frac{2b+1}{64} \right) = \sum_{r=0}^{31} \sum_{s=0}^{31} a_{rs} e^{2\pi i (\frac{ra}{32} + \frac{r(2b+1)}{64})}$$

and compute the "Gabor coefficients"

$$(32) \qquad a_{rs}, \quad 0 \leq r, s < 32,$$

by a two-dimensional 32 by 32 FFT. Strictly speaking these coefficients are periodized Gabor coefficients and already contain aliasing errors. These errors naturally occur in the sampling procedure. However, from these coefficients the samples of f at the points (29) can be synthesized exactly, up to machine error. Our problem is to analyze the information contained in varying size subsets of these coefficients. The greater the decay rate of the 32 by 32 array of Gabor coefficients, the more information is contained in a fixed size set of these coefficients. A related but important aspect of this analysis is to compute the information gain resulting from increasing the size of the coefficient set.

In fact, the largest L coefficients, $L = 12, 24, 48, 96$, were chosen for reconstruction. In case the cut off point eliminated some points of equal value, L was increased to include all samples of equal value. Reconstruction was accomplished by a two-dimensional 32 by 32 FFT, multiplication by the samples of $Z(g)$ and an inverse Zak transform, yielding f_{recon}. An error measure was constructing by computing

$$(33) \qquad \sum_s (f - f_{\text{recon}})^2$$

and normalizing by signal power. The results are presented in the tables below.

A quick glance at these tables shows four distinct classes verifying the results implied by the theorem. If $n \equiv 0 \bmod 4$, relatively few Gabor coefficients are required to reconstruct f to a high degree of accuracy. However, as $n \equiv 0 \bmod 4$ increases, more coefficients are required for the same degree of accuracy. For example, for a error of approximately .25%, 15 coefficients are required for H_4 and 27 coefficients are required for H_8. The next best case occurs when $n \equiv 3 \bmod 4$. Although more coefficients are required in this case as compared to the first case to synthesize up to the same degree of accuracy, the same phenomena occurs as $n \equiv 3 \bmod 4$ increases. The same remarks apply to the next best case $n \equiv 2 \bmod 4$. At the other extreme, $n \equiv 1 \bmod 4$, the error measure is relatively high even when $L \sim 96$.

For a fixed signal f and percentage error, the cost of carrying out Gabor computations digitally is measured by the number of Gabor coefficients required to synthesize f_{recon} approximating f, up to this percentage error. The four cases can be distinguished by the ratio of increased accuracy to increased cost. Whenever $n \equiv 0 \bmod 4$, doubling the number of coefficients increase the accuracy 10-fold. The effect of doubling the number of coefficients on the accuracy was described in the introduction. Perhaps the most significant feature of this list is that when $n \equiv 1 \bmod 4$, doubling the cost does not produce a doubling in accuracy. Gabor expansions, in this case, do not seem to be efficient. A modification will be suggested below and is under investigation. The first case $n \equiv 0 \bmod 4$, is probably as good as we can expect but modifications are possible in the remaining two cases.

H_4	# Gabor coefficients	% error
	15	.261
	27	.027
	51	.0032
	101	.00026

H_8	14	3.3
	27	.29
	51	.034
	103	.0026

H_{12}	15	25.49
	29	2.37
	59	.19
	110	.020

Increasing the sampling rate by 2 improves the accuracy of the synthesized signal ten-fold. It is reasonable to define functions in this class as narrow band Gabor functions.

H_1	12	38.39
	25	23.48
	54	17.47
	108	12.27

H_5	13	31.74
	28	24.34
	60	12.37
	117	09.29

H_9	14	54.91
	29	41.38
	58	34.51
	114	27.05

Increasing the sampling rate by two results in only a percentage increase in accuracy. The term wideband Gabor functions is perhaps appropriate.

H_2	16	4.02
	33	1.30
	60	0.52
	116	0.19

H_6	15	10.47
	28	3.93
	57	1.34
	114	0.45

H_{10}	13	28.91
	36	10.11
	66	4.39
	120	1.71

Doubling the sampling rate improves the accuracy 3-fold.

H_3	12	1.21
	28	.172
	57	.029
	110	.0055

H_7	13	6.10
	29	.86
	55	.175
	109	.029
H_{11}	15	2.001
	26	.67
	54	.15
	108	.03

Doubling the sampling rate produces a five-fold increase in accuracy.

4. Future Work. The result established for the Hermite functions H_4, H_8, H_{12} will be extended to a large library of Hermite functions $H_n, n \equiv 0 \mod 4$. It is expected that as $n \equiv 0 \mod 4$ increases, the number of coefficients required for a fixed percentage error will double as we go from n to $n + 4$. In addition, for a fixed $n \equiv 0 \mod 4$, we will test more completely the conjecture that doubling the coefficients increases accuracy ten fold. In general, this conjecture should hold for arbitrary linear combinations of Hermite functions of the form $H_n, n \equiv 0 \mod 4$. The case of any $\hat{f} = f$ is still open because the L^2- limit could produce surprises.

The case $\hat{f} = -if$ is already bad for the Hermite functions H_1, H_5, H_9. In this case, $Z(f), f = H_n, n \equiv 1 \mod 4$, vanishes at $(0,0)$. Expanding f as a modified Gabor expansion with $g(t) = e^{-\pi t^2}$ replaced by $g(t - \frac{1}{2})e^{\pi i t}$ promises better results since the Zak transform of the time-frequency shifted Gaussian also vanishes at $(0,0)$. Computer experiments are already underway. The remaining case will be subjected to similar analysis.

More generally, for smooth $Z(f)$, we have two ways of proceeding. We can decompose f into its eigenvectors relative to the Fourier transform and apply the above methods to each component. Further because $Z(f)$ must have a zero we can by time-frequency shift of the Gaussian, match any zero of $Z(f)$. This adoptive method should include some classification of the zeros of $Z(f)$. As seen in the theorem, not all zeros are equal.

REFERENCES

[1] L. AUSLANDER AND R. TOLIMIERI , *Abelian harmonic analysis, theta functions and function algebra on nilmanifolds*, Springer-Verlag, (1975).

[2] L. AUSLANDER AND R. TOLIMIERI, *Radar ambiguity function and group theory*, SIAM J. Math. Anal., 16(3), pp. 577–601, (1985).

[3] L. AUSLANDER AND R. TOLIMIERI, *Computing the decimated cross-ambiguity function*, IEEE Trans. on Acoustic, Speech, and Signal Processing, ASSP-36, 3, pp. 359–364, (1988).

[4] L. AUSLANDER AND R. TOLIMIERI, *On finite Gabor expansion of signals*, in *Signal Processing, Part I: Signal Processing Theory*, eds. L. Auslander, F.A. Grünbaum, J.W. Helton, T. Kailath, P. Khargonekar, S. Mitter, IMA Volumes in Mathematics and its Applcations, vol # 22, (1990).

[5]   L. AUSLANDER AND R. TOLIMIERI, *Time-Frequency Sampling Theory*, Preprint, 1990.

[6]   M. J. BASTIAANAS, *Gabor signal expansion and degrees of freedom of a signal*, Optica Acta, 29, pp. 1223–1229, (1982).

[7   D. GABOR, *Theory of communications*, J. IEE(London), 93, pp. 429-457, (1946).

[8]   A. J. E. M. JANSSEN, *The Zak transform: A signal transform for sampled time-continous signals* Philips J. Res., 43, pp. 23–69, (1988).

[9]   A. J. E. M. JANSSEN, *Gabor representation of generalized functions*, J. Math. Anal. Appl., Vol. 83, pp. 377–394, (1981.)

[10]   J-B. MARTENS, *The Hermite Transform-Theory and Applications*, IEEE Trans. on Acoustic, Speech, and Signal Processing, ASSP-38, No. 9, (1990).

[11]   W. SCHEMPP, *Radar Ambiguity Functions, Nilpotent Harmonic Analysis, and Holomorphic Theta Series, Special Functions: Group Theoretical Aspects and Applications*, D. Reidel Publishing, (1984).

[12]   R. TOLIMIERI, *Analysis on the Heisenberg manifold*, Trans. AMS, Vol. 228, pp. 329–343, (1977).

[13]   A. WEIL, *Sur certains groupes d'operatours unitaires*, Acta Math., 111, pp. 143–211, (1964).

[14]   J. ZAK, *Finite translations in a solid physics*, Phys. Res. lett., 19, pp. 1385–1397, 1967.

THE FINITE ZAK TRANSFORM AND
THE FINITE FOURIER TRANSFORM*

LOUIS AUSLANDER, IZIDOR GERTNER AND RICHARD TOLIMIERI[†]

Abstract. This paper presents the fundamental properties of the finite Zak transform. It is shown that the Zak transform appears as an intermediate result in the Cooley-Tukey FFT algorithm. The finite Zak transform is used to compute the Gabor expansion coefficients with sequentially increased resolution.

1. Introduction. The ultimate goal in signal processing is generally to achieve a processor output for which the desired signal stands out well above the accompanying undesirable natural or manmade noise. The purpose of this signal extraction is to estimate information-bearing parameters; to separate an information-carrying signal from "signal like" disturbances such as clutter or reverberation due to random scatterers in the signal propagation media in radar or sonar respectively; to resolve signals having similar characteristics such as, for example, close-spaced range and doppler radar or sonar echo signals. Such problems are encountered in applications such as seismology, communications, radar and sonar, image processing, biomedical signal processing.

Analysis of signals cannot always be accomplished by the simple use of classical time-domain representations such as correlation methods or frequency-domain representations based on the Fourier transform. The classical analysis is based on the stationarity assumption on the signals and Gaussian assumption on the interference. For signals not satisfying those assumptions, the concept of a time–frequency representation has been introduced.

Usually, time-frequency representations are bilinear transformations that display signal information over the euclidean plane. The Wigner-Ville distribution and, the ambiguity functions are the most common examples. In contrast, the Zak transform is a linear transformation that provides time-frequency information over the unit square. The Zak transform stands at center stage in the theory and application of Wigner distribution, ambiguity function, and Gabor expansions. In this work, the digital aspects of the theory will be considered.

The finite Zak transform (FZT) will be defined in the next section and the fundamental properties of the FZT will be described. Its relation to the Cooley-Tukey FFT algorithm will be shown. Finally the finite Zak transform, will be used to compute the Gabor expansion coefficients.

* This research was sponsored by Defense Advanced Research Projects Agency Darpa Order No.6674, Monitored by AFOSR Under Contract No.F49620-89-C-0020

† Center for Large Scale Computation, The Graduate School and University Center CUNY, 25 West 43rd Str. Suite 400, New York, N.Y. 10036

2. The Finite Zak Transform. Digital signal processing operates on finite data sets. For the purpose of simplicity and precision, we assume that the data is defined over the integers Z but subject to the following periodicity conditions. Choose a positive integer N . A complex-valued function $f(n)$, $n \in Z$, is called N–periodic if it satisfies

(1) $$f(n + N) = f(n), \quad n \in Z$$

An N– periodic function f is determined by its values

(2) $$f(0), f(1), \dots, f(N-1),$$

and will be called N–point data. Denote by $L^2(Z/N)$ the Hilbert space of all N–periodic functions with inner product

(3) $$\langle f, g \rangle = \sum_{n=0}^{N-1} f(n) g^*(n), \qquad f, g \in L^2(Z/N)$$

Suppose that N is not a prime and choose any factor L of N . *The finite Zak transform* $(Z(L)f)(a, b)$, $a, b \in Z$ of a function $f \in L^2(Z/N)$ is defined by the formula

(4) $$(Z(L)f)(a, b) = \sum_{r=0}^{L-1} f(a + Mr) e^{2\pi i \frac{br}{L}}, \quad N = L \cdot M.$$

Straightforward computation shows that $Z(L)f$ is N–periodic in each variable. In fact more is true.

THEOREM 1. *For $f \in L^2(Z/N)$, $N = L \cdot M$,*

(5) $$(Z(L)f)(a + M, b) = e^{-2\pi i \frac{b}{L}} (Z(L)f)(a, b),$$

(6) $$(Z(L)f)(a, b + L) = (Z(L)f)(a, b), \quad a, b \in Z.$$

Proof.

$$
\begin{aligned}
(Z(L)f)(a + M, b) &= \sum_{r=0}^{L-1} f(a + M + Mr) e^{2\pi i \frac{br}{L}} \\
&= \sum_{r=0}^{L-1} f(a + Mr) e^{2\pi i \frac{b(r-1)}{L}} \\
&= e^{-2\pi i \frac{b}{L}} (Z(L)f)(a, b)
\end{aligned}
$$

The second assertion follows directly from the L-periodicity of $e^{2\pi i \frac{rb}{L}}$ as a function of b . $\square$

By Theorem 1, $Z(L)f$ is completely determined by its values

(7) $$Z(L)f(a, b), \quad 0 \le a < M, \ 0 \le b < L.$$

For each $0 \le a < M$, we compute the L values

$$Z(L)f(a, b), \quad 0 \le b < L,$$

by computing the L –point Fourier transform $F(L)$ of the data set

$$(8) \qquad f(a), f(a+M), \ldots, f(a+(L-1)M).$$

Overall, computing $Z(L)f$ in this way requires M L–point Fourier transforms.

The finite Zak transform preserves information. The function $f(a)$, $0 \leq a < N$, can be recovered from its Zak transform by the formula

$$(9) \qquad f(a) = L^{-1} \sum_{b=0}^{L-1} (Z(L)f)(a,b), \quad ,0 \leq a < N.$$

For each $0 \leq a < N$ we view (4) as the Fourier series expansion in $L^2(Z/L)$ of $Z(L)f(a,b)$, $0 \leq b < L$. The Fourier coefficients are given by the data set (8). To derive an inversion formula which depends solely on the values of the finite Zak transform given in (7), we observe that as a runs over $0 \leq a < M$, the data set (8) completely describes f. By Fourier inversion we have

$$(10) \qquad f(a+Mr) = L^{-1} \sum_{b=0}^{L-1} (Z(L)f)(a,b)e^{-2\pi i \frac{br}{L}}, \quad 0 \leq a < M, \quad 0 \leq r < L.$$

Conditions (5) and (6) of Theorem 1 completely characterize the space of finite Zak transforms corresponding to L.

THEOREM 2. *Suppose $F(a,b)$, $a, b \in Z$ is a complex-valued function satisfying*

$$(11) \qquad F(a+M, b) = e^{-2\pi i \frac{b}{L}} F(a,b),$$

$$(12) \qquad F(a,b+L) = F(a,b), \quad a, b \in Z.$$

for some positive integer L and M.

Then the function

$$(13) \qquad f(a) = L^{-1} \sum_{b=0}^{L-1} F(a,b), \quad a \in Z,$$

is in $L^2(Z/N)$, $N = L \cdot M$, and

$$(14) \qquad F(a,b) = Z(L)f(a,b), \quad a, b \in Z.$$

Proof. Condition (11) implies F is N–periodic in the variable a:

$$
\begin{aligned}
F(a+N, b) &= e^{-2\pi i \frac{b}{L}} F(a+(L-1)M, b) \\
&= e^{-2\pi i \frac{2b}{L}} F(a+(L-2)M, b) \\
&\;\;\vdots \\
&= e^{-2\pi i \frac{Lb}{L}} F(a,b) \\
&= F(a,b)
\end{aligned}
$$

Consequently, f is N–periodic. Because F and $Z(L)f$ satisfy conditions (11) and (12), we need to prove (14) solely for $0 \leq a < M$, $0 \leq b < L$. Applying (13) to

$$Z(L)f(a,b) = \sum_{r=0}^{L-1} f(a + Mr)e^{2\pi i \frac{rb}{L}}$$

$$= L^{-1} \sum_{r=0}^{L-1}\sum_{c=0}^{L-1} F(a + rM, c)e^{2\pi i \frac{rb}{L}}$$

we have, for $0 \leq a < M$, $0 \leq b < L$, by (11)

$$(Z(L)f)(a,b) = L^{-1} \sum_{c=0}^{L-1} F(a, c) \sum_{r=0}^{L-1} e^{2\pi i \frac{r(b-c)}{L}}$$

$$= F(a, b)$$

since in the case

$$L^{-1} \sum_{r=0}^{L-1} e^{2\pi i \frac{r(b-c)}{L}} = \begin{cases} 1 & , \quad b = c, \\ 0 & , \quad otherwise. \end{cases}$$

□

Denote by $L^2(Z(L))$ the Hilbert space of all complex-valued functions $F(a,b)$, $a, b \in Z$, satisfying conditions (11) and (12) with inner product defined by

(15) $$\langle F, G \rangle = \sum_{a=0}^{M-1}\sum_{b=0}^{L-1} F(a, b)G^*(a, b).$$

THEOREM 3. *The mapping $L^{-\frac{1}{2}}Z(L)$ is an isometry from $L^2(Z/N)$ onto $L^2(Z(L))$: If $f \in L^2(Z/N)$ and $F = \mathbb{Z}(L)f$ then*

(16) $$\| f \|^2 = L^{-1} \| Z(L)f \|^2 = L^{-1} \| F \|^2 .$$

Proof.

$$\| F \|^2 = \langle F, F \rangle$$

$$= \sum_{a=0}^{M-1}\sum_{b=0}^{L-1}\sum_{r=0}^{L-1}\sum_{s=0}^{L-1} f(a + rM)f^*(a + sM)e^{2\pi i \frac{(r-a)b}{L}}$$

$$= \sum_{a=0}^{M-1}\sum_{r=0}^{L-1}\sum_{s=0}^{L-1} f(a + rM)f^*(a + sM) \sum_{b=0}^{L-1} e^{2\pi i \frac{(r-a)b}{L}}$$

$$= \sum_{a=0}^{M-1}\sum_{r=0}^{L-1} | f(a + rM) |^2 \cdot L$$

As a runs over $0 \leq a < M$ and r runs over $0 \leq r < L$, we have $a + rM$ running over $0 \leq a + rM < N$. Consequently

$$\| F \|_2^2 = L \cdot \| f \|^2,$$

proving the theorem. □

3. The Finite Fourier Transform and Cooley-Tukey FFT. The finite Zak transform is, in fact, the main building block for the Cooley-Tukey FFT algorithm [18]. First we prove the following fundamental result.

THEOREM 4. *If $f \in L^2(Z/N)$ and $\hat{f}$ is its N-point Fourier transform then*

$$(17) \qquad e^{-2\pi i \frac{ab}{N}}(Z(L)f)(-b, a) = M^{-1}\left(Z(M)\hat{f}\right)(a, b)$$

Proof. Set
$$F(a, b) = e^{-2\pi i \frac{ab}{N}}(Z(L)f)(-b, a).$$
Direct computation shows that F satisfies conditions (11) and (12) with M and L interchanged
$$F(a + L, b) = e^{-2\pi i \frac{b}{M}} F(a, b)$$
$$F(a, b + M) = F(a, b).$$

By Theorem 2
$$F(a, b) = (Z(M)g)(a, b)$$
where
$$g(a) = M^{-1} \sum_{b=0}^{M-1} F(a, b), \quad a \in Z.$$

Expanding
$$
\begin{aligned}
g(a) &= M^{-1} \sum_{b=0}^{M-1} \sum_{r=0}^{L-1} f(-b + Mr) e^{-2\pi i \frac{a}{N}(-b+Mr)} \\
&= M^{-1} \sum_{n=0}^{N-1} f(n) e^{-2\pi i \frac{an}{N}} \\
&= M^{-1} \hat{f}(a),
\end{aligned}
$$

proving the theorem. □

Theorem 4 is crucial to understanding the role of the finite Zak transform in digital signal processing relating to the Fourier transform. Loosely speaking, formula (17) asserts that, up to the factor $e^{-2\pi i \frac{ab}{N}}$, 90° rotation of the Zak transform $Z(L)f$ produces the Zak transform $Z(M)\hat{f}$.

The main part of Cooley–Tukey FFT algorithm is also contained in formula (17). Neglecting permutations we can compute $\hat{f}$ by the following sequence of steps.

1. Compute $(Z(L)f)(a, b)$, $0 \le a < M$, $0 \le b < L$. This step requires M L-point Fourier transforms.

2. Twiddle by the factor $e^{-2\pi i \frac{ab}{N}}$. This step requires N multiplications as $0 \le a < N$, $0 \le b < L$.

3. Compute the inverse Zak transform $Z(M)^{-1}$ by (10) (with L and M interchanged). This step requires L M-point Fourier transforms.

4. Wavelets. Let $g \in L^2(Z/N)$ and define for $m, n \in Z/N$, the function $g_{mn} \in L^2(Z/N)$ by the formula

(18) $$g_{mn}(a) = g(a+m)e^{2\pi i \frac{na}{N}}, \quad a \in Z/N.$$

The functions $g_{mn}(a)$, $m, n \in Z/N$, are called *finite Heisenberg group wavelets* [2] or simply *wavelets*. For fixed g, we will use the finite Zak transform to study properties of the wavelets corresponding to g. Take $N = L \cdot M$.

THEOREM 5. *For* $m, n \in Z/N$,

(19) $$(Z(L)g_{mn})(a,b) = e^{-2\pi i \frac{an}{N}}(Z(L)g)(a+m, b-n).$$

Proof. The left side of (19) is

$$\sum_{r=0}^{L-1} g_{mn}(a+rM)e^{2\pi i \frac{rb}{L}} =$$

$$= \sum_{r=0}^{L-1} g(a+m+rM)e^{-2\pi i \frac{n(a+rM)}{N}} \cdot e^{2\pi i \frac{rb}{L}}$$

$$= e^{-2\pi i \frac{na}{N}} \sum_{r=0}^{L-1} g(a+m+rM)e^{2\pi i \frac{r(b-n)}{N}M}$$

$$= e^{-2\pi i \frac{na}{N}}(Z(L)g)(a+m, b-n),$$

proving the theorem. □

COROLLARY 1. *For* $0 \leq m' < L$, $0 \leq n' < M$,

(20) $$(Z(L)g_{m'M, n'L})(a,b) = (Z(L)g)(a,b)e^{-2\pi i(\frac{n'a}{M}+\frac{m'b}{L})}$$

Proof. By Theorem 5, the left side of (20) is

$$e^{-2\pi i \frac{an'}{M}}(Z(L)g)(a+m'M, b-n'L)$$

which by Theorem 1 is

$$e^{-2\pi i \frac{an'}{M}}(Z(L)g)(a+m'M, b)$$

$$= e^{-2\pi i(\frac{n'a}{M}+\frac{m'b}{L})}(Z(L)g)(a,b),$$

proving the corollary. □

One of the main applications of (20) will be to replace certain (nonlinear) computations in part by a two-dimensional finite Fourier transform. First we will determine necessary and sufficient conditions for a wavelet basis of $L^2(Z/N)$.

THEOREM 6. *The set*

(21) $$\{g_{m'M, n'L} : 0 \leq m' < L, 0 \leq n' < M\}$$

is a basis of $L^2(Z/N)$ if and only if $Z(L)g$ never vanishes.

Proof. Suppose for all $0 \le m' < L$, $0 \le n' < M$,

$$\langle f, g_{m'M,n'L} \rangle = 0.$$

By Theorem 2 and the preceeding corollary

$$\langle Z(L)f, Z(L)g_{m'M,n'L} \rangle = 0$$

which can be rewritten as

$$\sum_{a=0}^{M-1} \sum_{b=0}^{L-1} (Z(L)f)(a,b)(Z^\star(L)g)(a,b) e^{-2\pi i (\frac{na'}{M} + \frac{m'b}{L})} = 0.$$

The double summation is a two-dimensional $M \times L$ Fourier transform. It follows that the products

$$Z(L)F(a,b)Z^*(L)g(a,b) = 0, \qquad 0 \le a < M, 0 \le b < L.$$

By Theorem 1 this holds for all $a, b \in Z$.

If $Z(L)g$ never vanishes then $Z(L)f$ is identically zero. Consequently, f is identically zero and the wavelets (21) span $L^2(Z/N)$. By dimension they form a basis of $L^2(Z/N)$

Conversely, if $Z(L)g$ vanishes somewhere than a nontrivial f can be found such that all products $(Z(L)f)(a,b)(Z^*(L)g)(a,b) = 0$. Reversing the steps of the preceeding argument, f is a nontrivial function not contained in the linear span of the wavelets in (21) implying this wavelet set is not a spanning set, completing the proof of the theorem. $\square$

The product function

$$(22) \qquad (Z(L)f)(a,b)\,(Z^*(L)g)\,(a,b), \qquad a, b \in Z,$$

is M-periodic in the variable a and L-periodic in the variable b, by Theorem 1. From the proof of Theorem 6,

$$(23) \qquad (Z(L)f)\,(a,b)\,(Z^*(L)g)\,(a,b) =$$
$$= \sum_{m'=0}^{L-1} \sum_{n'=0}^{M-1} \langle f, g_{m'M,n'L} \rangle e^{2\pi i (\frac{n'a}{M} + \frac{m'b}{L})}$$

which expresses the product (22) as the inverse two-dimensional $M \times L$ Fourier transform of the array

$$(24) \qquad [\langle f, g_{m'M,n'L} \rangle]_{0 \le n' < M, 0 \le m' < L} .$$

These results will be 'shifted' to other wavelet sets. Take

$$(25) \qquad m = m'' + m'M, \qquad 0 \le m'' < M, 0 \le m' < L,$$
$$(26) \qquad n = n'' + n'L, \qquad 0 \le n" < L, 0 \le n' < M,$$

in the following.

Direct computation shows that

(27)
$$g_{mn}(a) = e^{2\pi i \frac{n''m'}{L}} \cdot (g_{m'',n''})_{m'M,n'L}(a)$$

Replacing g by $g_{m'',n''}$ in the preceeding discussion, we have the following.

THEOREM 7. *For $m, n \in Z/N$ as in (25) (26),*

(28)
$$(Z(L)g_{mn})(a,b) = e^{2\pi i \frac{n''m'}{L}} (Z(L)g_{m'',n''})(a,b)e^{2\pi i(\frac{n'a}{M} + \frac{m'b}{L})}$$

Proof. Arguing as above, the Fourier series expansion of the product function

(29)
$$(Z(L)f)(a,b)(Z^\star(L)g_{m'',n''})(a,b),$$

viewed as M-periodic in the variable a and L-periodic in the variable b is

(30)
$$\sum_{m'}^{L-1} \sum_{n'=0}^{M-1} \langle f, (g_{m'',n''})_{m'M,n'L} \rangle e^{2\pi i\left(\frac{n'a}{M} + \frac{m'b}{L}\right)}.$$

It follows that for fixed $0 \le m'' < M$, $0 \le n'' < L$, the set of wavelets

(31)
$$\{g_{m''+m'M,n''+n'L} : 0 \le m' < L, \qquad 0 \le n' < M\}$$

is a basis of $L^2(Z/N)$ *if and only if* $(Z(L)g_{m'',n''})$ never vanishes. In light of Theorem 5, this is equivalent to $Z(L)g$ never vanishing.

It is always the case that for nontrivial g, the set of wavelets

(32)
$$\{g_{mn} : 0 \le m, n < N\}$$

spans $L^2(Z/N)$. To see this suppose

$$\langle f, g_{mn} \rangle = 0 \text{ for all } 0 \le m, n < N.$$

Then each of the product functions (30) vanishes for

$$0 \le m'' < M, 0 \le n'' < L.$$

This implies that $Z(L)f$ is identically zero unless for some

$$(a_0, b_0), 0 \le a_0 < M, 0 \le b_0 < L,$$

(33)
$$(Z(L)g_{m'',n''})(a_0, b_0) = 0, \qquad 0 \le m'' < M, 0 \le n'' < L.$$

Applying Theorem 6

(34)
$$(Z(L)g)(a_0 + m'', b_0 - n'') = 0, \quad 0 \le m'' < M, 0 \le n'' < L.$$

Theorem 1 implies $Z(L)g$ vanishes identically, proving the claim.

THEOREM 8. *The set of all wavelets corresponding to a nontrivial g spans $L^2(Z/N)$. If*

(35)
$$(m_j'', n_j''), \qquad 1 \le j \le r,$$

where $0 \leq m_j'' < M$, $0 \leq n_j'' < L$, *and if for all* (a,b) *satisfying* $0 \leq a < M$, $0 \leq b < L$,

(36) $$(Z(L)g_{m_j'',n_j''})(a,b) \neq 0$$

for some j then the set of wavelets

(37) $$\left\{ g_{m_j''+m'M,n_j''+n'L} : 1 \leq j \leq r,\ 0 \leq m' < L,\ 0 \leq n' < M \right\}$$

spans $L^2(Z/N)$.

5. Finite Cross-Ambiguity Function. The *finite cross-ambiguity function* of two functions $f, g \in L^2(Z/N)$ is the function $A(f,g)(m,n)$, $m, n \in Z$, defined by the formula

(38) $$A(f,g)(m,n) = \langle f, g_{mn} \rangle$$

The finite cross-ambiguity function is N-periodic in both variables. In [2] algorithms were designed to compute decimated values of the finite cross-ambiguity function in the case that m and n are powers of 2. We will extend these results in this section following [3].

First we will state without proof two elementary but important properties of $A(f,g)$.

THEOREM 9. *If* $f, g \in L^2(Z/N)$ *then*

(39) $$\|A(f,g)\|_2^2 = N\|f\|^2\,\|g\|^2$$

where

(40) $$\|A(f,g)\|_2^2 = \sum_{m=0}^{N-1}\sum_{n=0}^{N-1} |\,A(f,g)(m,n)\,|^2\,.$$

THEOREM 10. *If* f *and* g *are in* $L^2(Z/N)$ *and* $\hat{f}$ *and* $\hat{g}$ *are their Fourier transform then*

(41) $$A(f,g)(m,n) = e^{-2\pi i \frac{mn}{N}} A(\hat{f},\hat{g})(n,-m).$$

We can rewrite (23) as

(42) $$(Z(L)f)(a,b)(Z^*(L)g)(a,b) = \sum_{m'=0}^{L-1}\sum_{n'=0}^{M-1} A(f,g)(m'M,n'L)\,e^{2\pi i\left(\frac{n'a}{M}+\frac{m'b}{L}\right)}$$

It follows that the 'decimated' cross-ambiguity function

(43) $$A(f,g)(m'M,n'L), \qquad 0 \leq m' < L,\ 0 \leq n' < M,$$

can be computed by taking the $M \times L$ two-dimensional Fourier transform of the product function

(44) $$(Z(L)f)(a,b)(Z^*(L)g)(a,b), \quad 0 \leq a < M,\ 0 \leq b < L.$$

Neglecting data permutation, the following sequence of steps computes the decimated cross - ambiguity function.

1. Compute $(Z(L)f)(a,b)$ and $(Z^*(L)g)(a,b)$,

   for $0 \leq a < M$, $0 \leq b < L$. This step requires $2M$ L-point Fourier transforms.

2. Compute the product function

$$(Z(L)f)(a,b)(Z^*(L)g)(a,b), \qquad 0 \leq a < M, 0 \leq b < L.$$

   This step requires N multiplications.

3. Compute the $M \times L$ two-dimensional Fourier transform of the product function. This step can be carried out using M L-point and L M-point Fourier transforms.

In all, $3M$ L-point and L M-point Fourier transform and N multiplications are required.

Replacing g by $g_{m'',n''}$ in the above discussion leads to an algorithm computing shifted decimated values:

$$(45) \qquad A(f,g)(m'' + m'M, n'' + n'L), \qquad 0 \leq m' < L, 0 \leq n' < M.$$

First for fixed $0 \leq m'' < M$, $0 \leq n'' < L$, we use (27) to write

$$(46) \qquad A(f,g)(m'' + m'M, n'' + n'L) = e^{-2\pi i \frac{n''m'}{L}} A(f, g_{m'',n''})(m'M, n'L),$$

where $0 \leq m' < L$, $0 \leq n' < M$. We compute the decimated values

$$(47) \qquad A(f, g_{m'',n''})(m'M, n'L), \qquad 0 \leq m' < L, 0 \leq n' < M$$

by the proceeding algorithm replacing g by $g_{m'',n''}$. Multiplying this array (47) by

$$e^{-2\pi i \frac{n''m'}{L}}$$

completes the computation. In all, $3M$ L-point, L M-point and $2N$ multiplications are required.

In a typical application, g is fixed and certain computations can be precomputed. Suppose this is the case. Precomputing

$$Z(L)g_{m'',n''}, \qquad 0 \leq m'' < M, \ 0 \leq n'' < L,$$

we can compute the complete set of values

$A(f,g)(m,n), 0 \leq m, n < N$, by the following sequence of steps.

1. Compute $(Z(L)f)(a,b)$, $\qquad 0 \leq a < M, 0 \leq b < L$.

2. For each $0 \leq m'' < M$, $0 \leq n'' < L$, compute the $M \times L$ two-dimensional Fourier transform of the product function corresponding to $g_{m'',n''}$. This step requires NM L-point and LN M-point Fourier transforms.

3. Multiply each of the arrays by the factor

$$e^{-2\pi i \frac{n''m'}{L}}.$$

   This step requires N^2 multiplications.

In all, $(N+1)M$ L-point and $N \cdot L$ M-point Fourier transforms and $2N^2$ multiplications are required. Also, observe that steps 2-4 can be computed in parallel.

6. Finite Gabor Expansion. A finite Gabor expansion of f of $L^2(Z/N)$ is any expansion of f as a linear combination of wavelets :

$$(48) \qquad f = \sum_{m=0}^{N-1} \sum_{n=0}^{N-1} c(m,n) g_{mn}.$$

The coefficients $c(m,n)$, $0 \le m,n < N$, are called the Gabor coefficients of the expansion. Because the set of all wavelets is not linearly independent, f admits several Gabor expansions. In contrast, the inner product $\langle f, g_{mn} \rangle$, $0 \le m, n < N$, provides a unique measure of the occurence of g_{mn} in f. Following [2], we can view (48) as a time-frequency representation of f where g_{mn} describes a 'packet' of information at the point $(-m, -n)$ in the time-frequency plane $Z/N \times Z/N$ corresponding to a time-frequency shift of g.

Fix a factorization $N = L \cdot M$ and suppose $Z(L)g$ never vanishes. By Theorem 6, every $f \in L^2(Z/N)$ can be written *uniquely* in the form

$$(49) \qquad f = \sum_{m'=0}^{L-1} \sum_{n'=0}^{M-1} c(m'M, n'L) g_{m'M,n'L}$$

The coefficient $c(m'M, n'L)$, $0 \le m' < L$, $0 \le n' < M$, is called the Gabor coefficient of f relative to $g_{m'M,n'L}$. For a fixed factorization, uniqueness is forced!

Applying the finite Zak transform to both sides of (49), we have by theorem 5

$$(50) \qquad (Z(L)f)(a,b) = (Z(L)g)(a,b) \sum_{m'=0}^{L-1} \sum_{n'=0}^{M-1} c(m'M, n'L) e^{-2\pi i(\frac{an'}{M} + \frac{bm'}{L})}.$$

The Gabor coefficients of f relative to the basis $g_{m'M,n'L}$, $0 \le m' < L$, $0 \le n' < M$,

$$(51) \qquad c(m'M, n'L), \qquad 0 \le m' < L, \ 0 \le n' < M,$$

can be computed by the inverse $M \times L$ two-dimensional Fourier transform of the quotient function

$$(52) \qquad \frac{(Z(L)f)(a,b)}{(Z(L)g)(a,b)}, \qquad 0 \le a < M, \ 0 \le b < L.$$

The Gabor coefficients (51) correspond to the lattice $MZ/N \times LZ/N$ of $Z/N \times Z/N$. Gabor coefficients corresponding to shifted lattices can be defined in an analogous manner and provide a finer time-frequency resolution.

For $0 \le m'' < M$, $0 \le n'' < L$. Every $f \in L^2(Z/N)$ has a unique expansion

$$(53) \qquad f = \sum_{m'=0}^{L-1} \sum_{n'=0}^{M-1} c(m'' + m'M, n'' + n'L) g_{m''+m'M,n''+n'L}$$

The inverse $M \times L$ two-dimensional Fourier transform of the quotient function

$$(54) \qquad \frac{(Z(L)f)(a,b)}{(Z(L)g_{m'',n''})(a,b)}, \qquad 0 \le a < M, \ 0 \le b < L$$

computes the Gabor coefficients of f relative to the basis

$$g_{m''+m'M,n''+n'L}, \qquad 0 \le m' < L, \ 0 \le n' < M.$$

Gabor coefficient and finite cross–ambiguity function values are related by two-dimensional convolution. By (3)

$$
(Z(L)f)(a,b)\overline{(Z^\star(L)g)(a,b)}
$$

(55)
$$
= \ |(Z(L)g)(a,b)|^2 \cdot \sum_{m'=0}^{L-1} \sum_{n'=0}^{M-1} c(m'M, n'L) e^{-2\pi i(\frac{an'}{M} + \frac{bm'}{L})}
$$

apply (23)

$$
(Z(L)f)(a,b)\overline{(Z^\star(L)g)(a,b)} =
$$

(56)
$$
= \sum_{m'=0}^{L-1} \sum_{n'=0}^{M-1} \langle f, g_{m'M,n'L} \rangle e^{2\pi i(\frac{an'}{M} + \frac{bm'}{L})}.
$$

Since

(57)
$$
|(Z(L)g)(a,b)|^2 = \sum_{m'=0}^{L-1} \sum_{n'=0}^{M-1} \langle g, g_{m'M,n'L} \rangle e^{2\pi i(\frac{n'a}{M} + \frac{m'b}{L})}
$$

we have that the array

(58)
$$
\langle f, g_{m'M,-n'L} \rangle, \qquad 0 \le m' < L, \ 0 \le n' < M
$$

can be written as the $L \times M$ two-dimensional cyclic convolution of the arrays

(59)
$$
\langle g, g_{m'M,n'L} \rangle, \qquad 0 \le m' < L, \ 0 \le n' < M
$$
(60)
$$
c(m'M, n'L), \qquad 0 \le m' < L, \ 0 \le n' < M.
$$

7. Computation of Gabor Coefficients with Sequentially Increased Resolution.

Fix a factorization $N = L \cdot M$ and assume $Z(L)g$ never vanishes. Consider the basis of $L^2(Z/N)$

(61)
$$
g_{m'M,n'L}, \qquad 0 \le m' < L, \ 0 \le n' < M,
$$

and also for each fixed $0 \le m'' < M, \ 0 \le n'' < L$, the basis of $L^2(Z/N)$

(62)
$$
g_{m''+m'M,n''+n'L}, \qquad 0 \le m' < L, \ 0 \le n' < M.
$$

We shall derive a formula relating these bases. This formula will be used to express finite ambiguity function values on the shifted lattice $(m'' + MZ) \times (n'' + LZ)$ in terms of finite ambiguity function values on the lattice $MZ \times LZ$.

A second application will relate Gabor coefficients with respect to these two bases.

Write

(63)
$$
g_{m''+m'M,n''n'L} = \sum_{r'=0}^{L-1} \sum_{s'=0}^{M-1} \alpha(m', n'; r', s') g_{r'M,s'L}.
$$

Arguing as in the preceeding section, the quotient function

$$(64) \qquad \frac{(Z(L)g_{m''+m'M,n''+n'L})\,(a,b)}{(Z(L)g)\,(a,b)}, \qquad 0 \le a < M, \ 0 \le b < L,$$

has Fourier series

$$(65) \qquad \sum_{r'=0}^{L-1} \sum_{s'=0}^{M-1} \alpha(m',n';r's')e^{-2\pi i(\frac{r'a}{M}+\frac{s'b}{L})}.$$

Apply Theorem 7, we can rewrite the quotient function as

$$(66) \qquad e^{2\pi i \frac{n''m'}{L}} G(a,b)e^{-2\pi i(\frac{n'a}{M}+\frac{m'b}{L})}$$

where

$$(67) \qquad G(a,b) = \frac{(Z(L)g_{m'',n''})\,(a,b)}{(Z(L)g)\,(a,b)}, \qquad 0 \le a < M, \ 0 \le b < L.$$

Throughout, we have suppressed the dependence of the Gabor coefficients and the function G on the choice of m'' and n''. Denote the $M \times L$ two-dimensional Fourier transform of G by $\widehat{G}$. Straightforward computation shows that

$$(68) \qquad \alpha(m',n';r's') = N^{-1}e^{2\pi i \frac{n''m'}{L}} \widehat{G}(n'-s',m'-s'),$$

proving the next result.

THEOREM 11. *For* $0 \le m'' < M$, $0 \le n'' < L$,

$$(69) \qquad g_{m''+m'M,n''+n'L} = e^{2\pi i \frac{m'n''}{L}} N^{-1} \sum_{r'=0}^{L-1} \sum_{s'=0}^{M-1} \widehat{G}(n'-s',m'-r')g_{r'M,s'L},$$

where G *is given in (67) and depends on* m'' *and* n''.

The double summation in (69) has the form of a two-dimensional cyclic convolution. Setting

$$(70) \qquad H(r',s') = \widehat{G}(s',r'), \qquad 0 \le r' < L, 0 \le s' < M,$$

we can rewrite the right side of (69) as

$$(71) \qquad e^{2\pi i \frac{m'n''}{L}} N^{-1} \sum_{r'=0}^{L-1} \sum_{s'=0}^{M-1} H(m'-r',n'-s')g_{r'M,s'L}.$$

The relationship between finite cross-ambiguity function values and finite Gabor expansions relative to the lattice $MZ \times LZ$ and shifts of this lattice is given in the next two corollaries.

COROLLARY 2. *For* $0 \le m'' < M$, $0 \le n'' < L$,

$$(72) \qquad A(f,g)(m''+m'M,n''+n'L)$$
$$= N^{-1}e^{-2\pi i \frac{m'n''}{L}} \sum_{r'=0}^{L-1} \sum_{s'=0}^{M-1} H^*(m'-r',n'-s')A(f,g)(r'M,s'L),$$

where H *is given in (70).*

COROLLARY 3. *If*

(73)
$$f = \sum_{m'=0}^{L-1} \sum_{n'=0}^{M-1} c(m'M, n'L) g_{m'M,n'L}$$

and

$$f = \sum_{m'=0}^{L-1} \sum_{n'=0}^{M-1} c(m'' + m'M, n'' + n'L) g_{m''+m'M,n''+n'L},$$

then

$$c(r'M, s'L)$$

(74)
$$= \quad N^{-1} \sum_{m'=0}^{L-1} \sum_{n'=0}^{M-1} c(m'' + m'M, n'' + n'L) H(m' - r', n' - s') e^{2\pi i \frac{m'n''}{L}},$$

where H is given in (70)

In both (72) and (74), the double series has the form of two-dimensional cyclic convolution. The presented formulas enable computation of the Gabor expansion coefficients with sequentially increasing resolution.

REFERENCES

[1] L.Auslander, I.Gertner, R.Tolimieri, *The discrete Zak transform application to time-frequency analysis and synthesis of non-stationary signals*, submitted for publication.

[2] L.Auslander,R.Tolimieri, *Computing decimated cross–ambiguity function* , IEEE Transactions on ASSP, vol.36, No.3, pp.359-364, March, 1988.

[3] L.Auslander, I.Gertner, *Computing the ambiguity function with sequentially increased resolution*, submitted for publication.

[4] L.Auslander, R.Tolimieri, *On finite Gabor expansion of signals*, in *Signal Processing, Part I: Signal Processing Theory*, eds. L. Auslander, F.A. Grünbaum, J.W. Helton, T. Kailath, P. Khargonekar, S. Mitter, IMA Volumes in Mathematics and its Applications, vol # 22, (1990).

[5] L.Auslander, I.Gertner, *Wide-band ambiguity function and ax+b group*, IMA Proceedings on Signal Processing, Minneapolis, 1988.

[6] L.Auslander, R.Tolimieri. *Abelian harmonic analysis, theta functions and function algebra on nilmanifolds*, Springer-Verlag, 1975.

[7] A.Weil, *Sur certains groupes d'operatours unitaires*, Acta Math. 111 , pp.143-211, 1964.

[8] R.Tolimieri, *Analysis on the Heisenberg manifold*, Transactions of the American Math.Soc., vol. 228, pp.329-343, 1977.

[9] A.J.E.M. Janssen, *The Zak transform : a signal transform for sampled time-continuous signals*, Philips J. Res. vol.43, pp.23-69, 1988.

[10] L.Auslander, R.Tolimieri, *Radar ambiguity function and group theory*, SIAM J.Math. Anal., vol.16, No.3, pp. 577-601, May, 1985.

[11] L.Auslander, R.Tolimieri, *Characterizing the radar ambiguity functions*, IEEE Trans. on Information Theory, vol. IT-30, No. 6, pp. 832-836, Nov. 1984.

[12] L.Auslander, R.Tolimieri, *Ambiguity functions and group representations* in " Group representations, Ergodic theory, Operator algebras, and Mathematical Physics" pp.1-10, Springer-Verlag, 1987.

[13] D.Gabor. *Theory of communications* , J.IEE(London) vol.93. pp.429-457 Nov 1946.

[14] J.Zak, *Finite Translations in Solid Physics,* Phys. Rev.Lett. 19(1967), 1385-1397.

[15] S.Key, F. Boudreaux-Bartels *On the optimality of the Wigner Distribution for detection,* pp.1017-1020, ICASP 1985.

[16] B.V.K. Vijava Kumar, C.W. Carroll *Performance of Wigner distribution function based detection methods,* Optical Engineering, No.6, vol.23, pp.732-737, 1984.

[17] I.Gertner *An Alternative Approach To Nonlinear Filtering,* Stochastic Processes and their Applications, 1976.

[18] J.W.Cooley, J.W.Tukey, *An Algorithm for the Machine Computation of Complex Fourier Series,* Math. Computation, Vol.19, pp. 297–301, Apr. 1965.

ON THE ALTERNATIVES FOR IMAGING ROTATIONAL TARGETS

MARVIN BERNFELD*

Abstract. The subject of this paper is the inverse problem with respect to electromagnetic scattering. When an apparent rotation is introduced into the radar-target model, (as for a rotating target) inverse synthetic-aperture radar techniques can be employed to enhance resolution in crossrange. Originally such systems incorporated the classical rectangularly-formatted, coherent, range-Doppler signal processing. The limitation of these systems, however, has led to the invention of alternative solutions. One of these has been called CHIRP Doppler radar, and it is currently being evaluated. Another alternative and the focal point of this paper is Walker's polar format. In this case, the backscattered signals yield the two-dimensional Fourier transform of the scattering distribution that is illuminated; inverting the Fourier transform subsequently will produce an image of the scattering distribution. It is shown that the CHIRP z-transform can provide an exact construction of the Fourier transform.

1. Introduction. The point-to-point distribution of microwave reflectivity (or point reflectivity) of rigid targets will normally supply topographical discriminants that can be used to recognize different target classes. The task of constructing a point-reflectivity image based on backscatter measurements is a problem in inverse electromagnetic scattering and the subject of this paper. When an apparent rotation is introduced into the radar-target model, inverse synthetic aperture radar (ISAR) techniques offer an approach to enhance the crossrange resolution. This is usually necessary to equalize resolution in range and crossrange because superior range resolution is normally available through special wide-bandwidth pulse-compression techniques.

This paper discusses an ISAR approach that is based on Walker's polar format idea for enhancing crossrange resolution in spotlight-mode synthetic-aperture radar (SAR) [1]. The discussion addresses the description of the individual backscattered signals during a sequence of pulse transmissions that are modulated by linear FM. One may observe that these signals resemble signals that have passed through the first stage of the CHIRP z-transform (see appendix). With this observation, it is possible to show that an exact application of Walker's polar-format idea can be achieved with signal processing that includes the CHIRP z-transform. In this use of the term "exact", the interpretation is derived from the perspective that previous discussions on this subject have considered an approximation of the backscattered signal which results in an inexact application of the polar-format imaging idea [2]. This matter is discussed in the text below.

2. Classical ISAR Systems. The original ISAR systems designs incorporated classical rectangularly-formatted, coherent, range-Doppler signal processing. In these systems, basically, the backscattered signals for a sequence of N pulse transmissions are pulse by pulse compressed and coherently combined. The associated signal processing format is rectangular: pulse compression is normally accomplished

---

*Raytheon Co., 430 Boston Post Road, Wayland, MA 01778. Supported by DARPA Order No. 7090 and Monitored by AFOSR Contract No. F49620–89–C–0116.

first, then the coherent integration is performed. This procedure yields an output that consists of a map of the point relectivity as a function of range and Doppler. Figure 1 illustrates these coordinates superimposed on the surface of a rotating spherical object; radar astronomy is a good example. Note the intersection of constant range and constant Doppler planes. The two surface points with the same range-Doppler coordinates are potentially ambiguous, but the ambiguity can be resolved by employing interferometry.

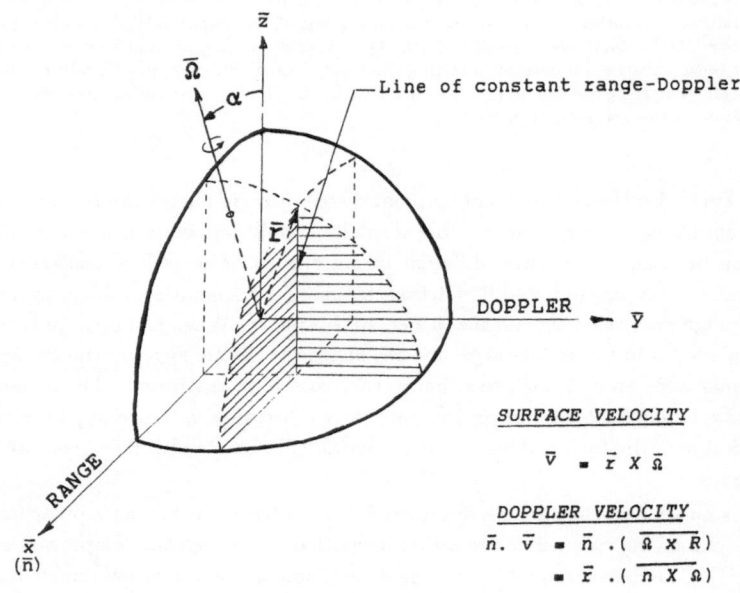

Figure 1   Range–Doppler Coordinate System of Rotating Sphere

The time interval over which backscatter from a rotating target can be integrated effectively is limited. While more Doppler resolution can be attained by extending the coherent integration, progressively more blurring will simultaneously be caused by range migration and diminishing Doppler coherence. The bound on effective coherent integration and, consequently, image quality is established when increased blurring overshadows the improvement in resolution. If, as a criteria for effective coherent integration, the movement of any peripheral scatterer is confined to a given range-Doppler resolution cell, then it can be shown this bound is a function of the target dimensions in range (D_r) and crossrange (D_a). The bound is expressed by two inequalities that define the smallest resolution cell meeting this criteria. These inequalities [3] are stated below, where ρ_r and ρ_a respectively, are the dimensions of the resolution cell in the range and crossrange directions, and λ_0 is the wavelength.

$$\rho_a^2 > \frac{\lambda_0 D_r}{4}$$

$$\rho_r \rho_a > \frac{\lambda_0 D_a}{4}$$

3. CHIRP Doppler Radar [4]. The limitation of the classical SAR systems with respect to image quality has led to additional SAR processing solutions. One of these has been called CHIRP Doppler radar; this is currently being analysed to prove that it can be a valuable imaging alternative. CHIRP Doppler radar attempts to resolve the limitation of the classical SAR by obtaining a snapshot of the target reflectivity density in range-Doppler coordinates. This is accomplished by exploiting the ridge-like ambiguity function that is characteristic of linear FM pulse-compression systems. Projections of a range-Doppler distribution are obtained and combined tomographically to form an image.

4. Polar Format Processing. This approach was inspired by the fact that after suitable demodulation of linear FM pulse echos, the backscatter per pulse can be treated as a radial profile of the two-dimensional Fourier transform of a scattering distribution. Walker was the first to recognize this relationship and initially applied the concept in SAR processing systems that incorporate optical signal processing. After a sufficient number of polar profiles are collected, as determined by resolution considerations, the inverse Fourier transform will yield an image of the target.

Figure 2 gives a simple model that can be employed for an explanation of the rpoceeding Fourier transform relationship. The model consists of a rotating free-form area that is co-planar with a remote CW radar. A distribution of microwave reflectivity is assumed to exist inside the boundary of the rotating area. This is denoted here by $\sigma(r, \theta)$ where r and θ are polar coordinates with respect to x, y cartesian coordinates centered at the axis aroound which the area is rotating. The range to the axis of rotation is denoted by R_0. It is assumed that this is known. In addition, it is assumed that R_0 is much greater than the maximum radial dimension $r_{\max}$ of the rotating area such that the following approximation of the range to any point in the area is valid:

$$R = R_0 + r \sin\{\theta - \phi\}$$

where ϕ denotes the angular displacement of the x and y, respectively, with respect to the fixed coordinates u and v.

The following additional constraints are included in the description of this model. It is assumed that the antenna beam covers the entire rotating area and that this area is uniformly illuminated by the radar's transmission. Also, it is assumed that the reflectivity at every point of the rotating area is isotrophic.

The composite stready state reflection from the rotating area in the model above, assuming that the angle of rotation is fixed while this area is illuminated by the radar, can be represented by the following echo-signal description.

$$\psi_R(t, \phi) = F\left(\frac{4\pi}{\lambda_0}, \phi\right) \exp\left\{-j\omega_0 \left(t - \frac{2R_0}{C}\right)\right\}$$

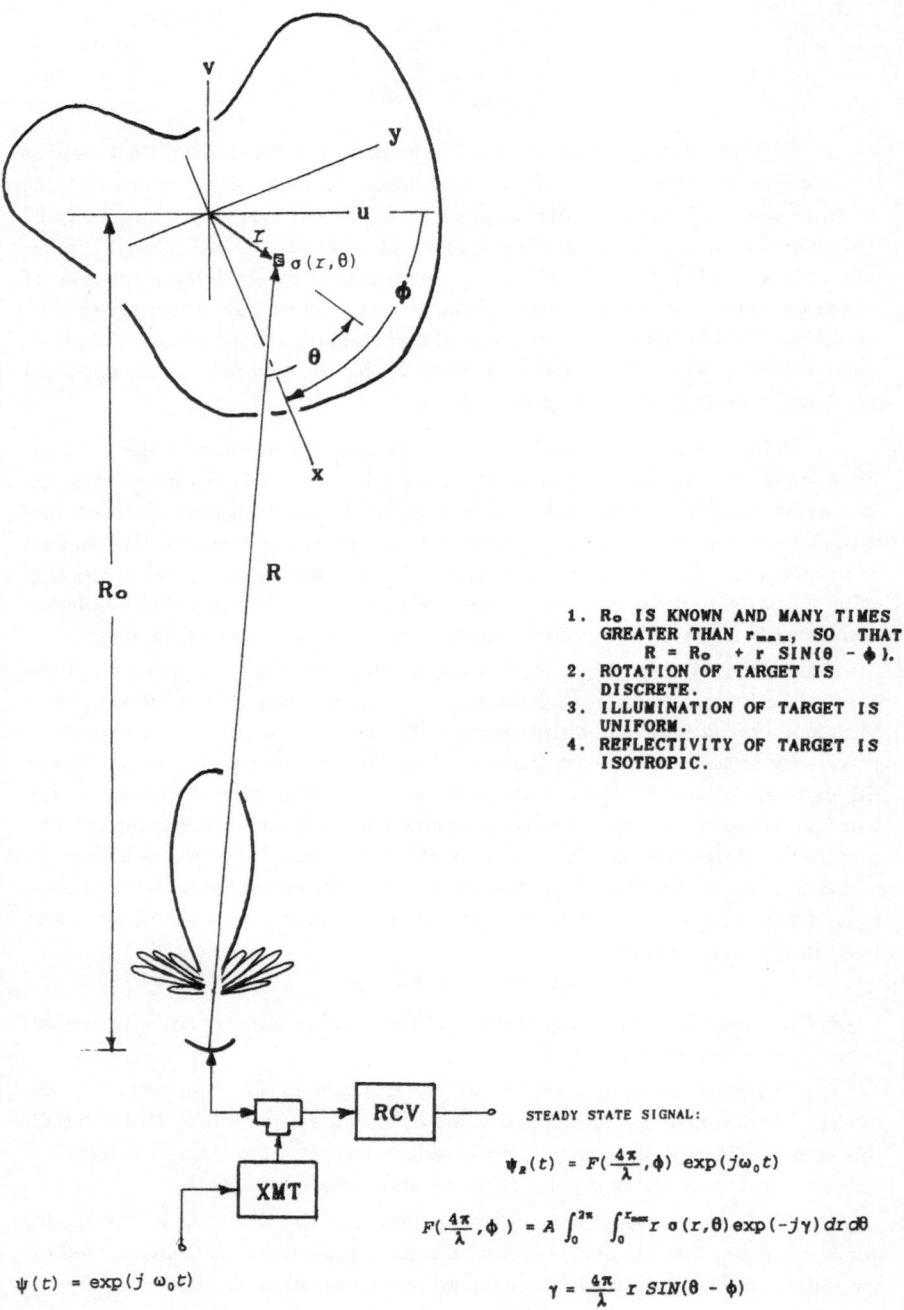

1. R_0 IS KNOWN AND MANY TIMES GREATER THAN r_{max}, SO THAT:
   $$R = R_0 + r\,SIN(\theta - \phi).$$
2. ROTATION OF TARGET IS DISCRETE.
3. ILLUMINATION OF TARGET IS UNIFORM.
4. REFLECTIVITY OF TARGET IS ISOTROPIC.

STEADY STATE SIGNAL:

$$\psi_R(t) = F(\tfrac{4\pi}{\lambda},\phi)\ \exp(j\omega_0 t)$$

$$F(\tfrac{4\pi}{\lambda},\phi) = A \int_0^{2\pi} \int_0^{r_{max}} r\,\sigma(r,\theta)\exp(-j\gamma)\,dr\,d\theta$$

$$\gamma = \frac{4\pi}{\lambda}\ r\,SIN(\theta - \phi)$$

$$\psi(t) = \exp(j\,\omega_0 t)$$

Figure 2   Radar-Target Model

where,

$$F\left(\frac{4\pi}{\lambda_0}, \phi\right) = A \int_0^{2\pi} \int_0^{r_m} r\sigma(r,\theta) \exp\{-j\gamma(\phi_i)\} dr\, d\theta$$

and,

$$\gamma(\phi) = \frac{4\pi}{\lambda_0} r \sin(\theta - \phi)$$

The amplitude of of this signal, denoted in the description above by $F(4\pi/\lambda_0, \phi)$ is equal to a point on the two-dimensional Fourier transform corresponding to $\sigma(\mathbf{r}, \theta)$; a circular locus of Fourier transforms with radius $4\pi/\lambda_0$ is obtained for a rotation of 360 degrees.

The CW transmission $\psi(t) = \exp\{j\omega_0 t\}$ assumed above can be replaced in the model by a linear FM pulse that is expressed as follows:

$$\psi(t) = \begin{cases} \exp\left\{-j\left(\omega t - \frac{1}{2}at^2\right)\right\} & ; \quad -\frac{T}{2} \leq t \leq \frac{T}{2} \\ 0 & ; \quad \text{elsewhere} \end{cases}$$

where $a = 2\pi\Delta f/T$, and, Δf and T, respectively, denote the swept bandwidth and the time duration of the pulse. It is assumed that the pulse length in spatial units ($cT/2$; c = light velocity) is many times the maximum radial dimension of the rotating area. In this case, the signal at the receiver input becomes:

$$\psi_R(t,\phi) = F(t,\phi) \exp\left\{-j\left[\omega_0\left(t - \frac{2R_0}{c}\right) - \frac{1}{2}a\left(t - \frac{2R_0}{c}\right)^2\right]\right\}$$

where

$$F(t,\phi) = A \int_0^{2\pi} \int_0^{\infty} r\sigma(r,\theta) \exp\{-j\gamma(t,\phi)\} dr\, d\theta$$

and,

$$\gamma(t,\phi) = \left[\frac{4\pi}{\lambda_0} + \frac{2}{c}a\left(t - \frac{2R_0}{c}\right)\right] r \sin(\theta - \phi) - \frac{a}{2}\left(\frac{2}{c}\right)^2 (r\sin(\theta - \phi))^2$$

$$; -\frac{T}{2} \leq t \leq \frac{T}{2}$$

The constraint on time in the definition of $\gamma(t,\phi)$ translates into a similar constraint on $F(t,\phi)$ that results ultimately in point-spreading in the reconstructed image. A point scatterer at the center of the rotating area, will yield essentially a constant amplitude annulus in the two-dimensional Fourier transform domain, as illustrated in Figure 3. For a relatively narrow annulus with respect to $4\pi/\lambda_0$, the point spreading is represented by an axial symmetric surface that is generated by a zero-order Bessel function.

Note the phase function, $\gamma(t,\phi)$, in the preceding description of the backscattered signal contains a quadratic term. If not for this term, Walker's polar format could be applied directly after the demodulation of the backscattered signal since by omitting this quadratic phase term, $F(t,\phi)$ equals the two-dimensional Fourier transform of the point reflectivity; the equality is confined to a band in the transform domain that depends on the bandwidth of the linear FM.

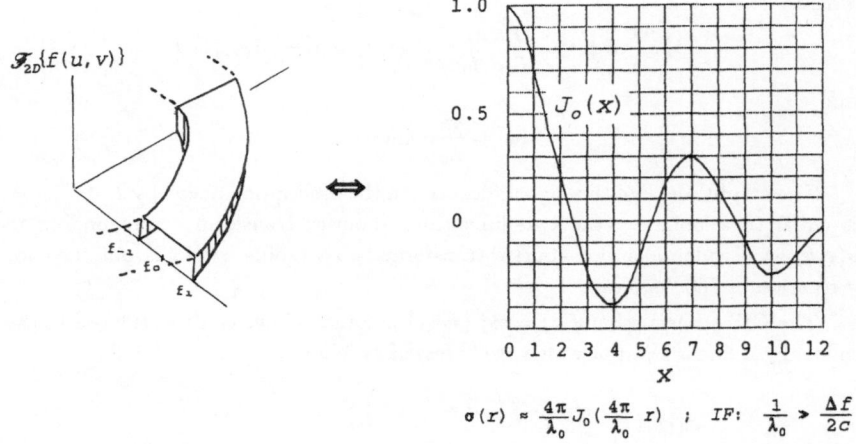

$$\sigma(r) \approx \frac{4\pi}{\lambda_0} J_0(\frac{4\pi}{\lambda_0} r) \quad ; \quad IF: \quad \frac{1}{\lambda_0} > \frac{\Delta f}{2c}$$

Figure 3   Point-Spread Function for Polar Format Imaging

It has been proposed that the quadratic-phase term can be ignored provided the maximum radial dimension of the rotating area does not exceed the size defined by the following [2]:

$$\frac{2r_m}{C} \leq \frac{1}{\Delta f} \left(\frac{T\Delta f}{2} \right)^{\frac{1}{2}}$$

This constraint of the maximum radial dimension is equivalent to the criteria that the phase error due to the omission of the quadratic phase term not exceed $\pi/2$.

In most cases the specification regarding the duration of the transmitted pulses, will ensure that the preceding criteria is fulfilled. Nevertheless, with respect to the signal processing design, it is not necessary to sacrifice any accuracy for the sake of simplicity gained by an approximation of the backscattered signal. Indeed, the complexity of the signal-processing design will not increase substantially when the quadratic-phase term is retained in the exact description of the phase function $\gamma(t\phi)$. This will demonstrated in the following paragraph.

Transforming to cartesian coordinates (u, v), will yield the following alternative description for $F(t, \phi)$.

$$F(t, \phi) = \int_{-\infty}^{\infty} p_\phi(v) \exp\left\{ j\frac{\varepsilon}{2} v^2 \right\} \exp\{-j2\pi\beta(t)v\} dv$$

where

$$1. \qquad \varepsilon = a \left(\frac{2}{c} \right)^2$$

$$2. \qquad \beta(t) = \left[\frac{2}{\lambda_0} + \frac{a}{2\pi} \left(\frac{2}{c} \right) \left(t - \frac{2R_0}{c} \right) \right] \; ; \; -\frac{T}{2} \le t \le \frac{T}{2}$$

Note the integrand in the preceding equation. This contains the factor $p_\phi(v)$, representing the Radon transform of the point reflectivity $\sigma(r, \theta)$ for a projection angle ϕ. It is expressed as follows:

$$p_\phi(v) = \int_{-\infty}^{\infty} f_\phi(u, v) du$$

where $f_\phi(u, v)$ is a ϕ-rotate of $\sigma(r, \phi)$.

The alternative description for $F(t, \phi)$ can be viewed from the perspective that it resembles signals that have passed through the first stage of a CHIRP z-transform. In this case, the signal is the line-integral projection of the point reflectivity $\sigma(r, \theta)$ for an angular direction ϕ. To obtain the Fourier transform of this projection, and, subsequently, to apply Walker's polar format idea, it is necessary to complete the missing stages of the CHIRP z-transform. As illustrated in Figure 4, this consists of multiplying $F(t, \phi)$ by a quadratic-phase function and invoking the inverse Fourier transform, then postmultiplying by a quadratic-phase function. Incorporating these signal-processing stages does not substantially increase the design complexity

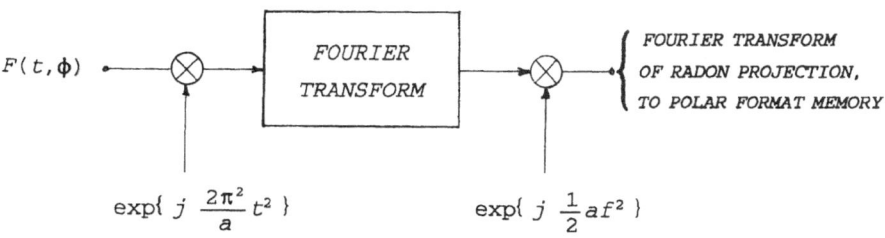

Figure 4 Polar Format – Signal Processing Format

5. Conclusion. By incorporating the missing stages that complete the CHIRP z-transform depicted in Figure 4 in the signal processing of the input backscatter, the image of a rotational object can be constructed by employing Walker's polar format idea based on the exact description of this backscatter signal. The complexity of the signal processing appears not to be substantially greater than when the backscattered signal is approximated by ignoring the quadratic-phase term in its exact description.

6. Appendix: CHIRP z-Transform [5]. The CHIRP-z transform is a computational algorithm for numerically evaluating the z-transform of a sequence of N samples. The z-transform is efficiently evaluated at M points in the z-plane which lie on circular or spiral contours beginning at any arbitrary point. The efficiency is derived from the fact that the values of the z-transform on a spiral or circular contour can be expressed as a discrete convolution. Thus, well-known high-speed convolution techniques can be applied to evaluate the transform efficiently. By performing analogous processing for analog input signals, as the schematic in Figure 5 indicates, the algorithm offers a convenient technique to obtain real-time Fourier transforms.

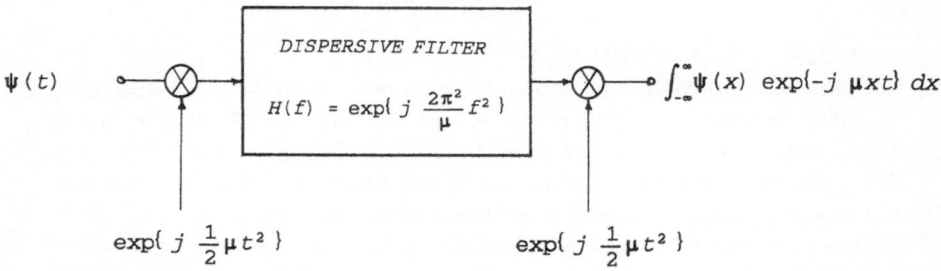

Figure 5   CHIRP z-Transform

REFERENCES

[1]   J.L. WALKER, *Range-Doppler imaging of rotating objects*, IEEE Trans. Aerosp. and Electron. Sys., AES–16 (Jan 1980), 23–52.

[2]   D.C. MUNSON, JR., J.D. O'BRIEN, K. JENKINS, *A tomographic formulation of Spotlight-Mode Synthetic Aperture Radar*, Proc. IEEE, 71 (Aug. 1983), 917–925.

[3]   E.D. BANTA, *Limitations on SAR image area due to motion through resolution cells*, IEEE Trans. Aerosp. and Elecron. Syst., AES–22 (Nov. 1986), 799–803.

[4]   M. BERNFELD, *CHIRP Doppler Radar*, Proc. IEEE, 72 (Apr. 1984), 540–541.

[5]   L.R. RABINER, R.W. SCHARFER, C.M. RADER, *The Chirp z-Transform algorithm and its application*, The Bell System Technical Journal (May–June 1969), 1249–1291.

PHASE MONOPULSE TRACKING AND ITS RELATIONSHIP TO NONCOOPERATIVE TARGET RECOGNITION

BRETT BORDEN*

Abstract. We examine the glint problem of radar tracking and show that a complete understanding of this problem is closely related to the problem of target identification. This relationship exists because it is the complex structure of the target that is fundamentally responsible for the difficulties in both tracking and noncooperative target recognition. In addition, these problems are further complicated by practical limitations in the radar data available from missile systems; they are narrowband, high frequency, and without detailed target aspect coordination.

1. Introduction. Much current research effort in small-missile system technology is directed toward imbuing future-generation missiles with greater tracking performance and on-board target identification ability. Ideally, the tracking and identification capabilities employed by these missile systems should function at ranges well beyond visual range (hundreds of kilometers), should operate in any weather, and must not require cooperation from a friendly target. Radar–based systems alone have the potential to satisfy these conditions.

Tracking schemes based upon *phase monopulse* detection have been used for many years [1]. Constant-phase surfaces of the scattered field, known as *phase fronts*, typically surround the scattering target. The gradient of phase will be orthogonal to these phase fronts and so can be expected to point in the direction of the target when the phase fronts are spherical. For tracking purposes, interferometric methods are used to determine phase front normals and this information is used to estimate target location. However, as a consequence of the mutual interference interaction between the individual scattering centers associated with complex scattering bodies, the phase fronts typically will have sufficient structure that their normals do not point exactly at the object itself. Often the implied tracking direction is *significantly* away from the actual target location. This is known as the *glint problem*, and the difference between the phase gradient and the true target direction is often called *glint*.

By comparison, target identification methods are not nearly as well developed as tracking methods. Despite impressive accomplishments in scattered-field based imaging (inverse scattering) in areas such as x-ray tomography and seismic prospecting, progress has been slow in applying inverse scattering to radar based airborne noncooperative-target-recognition schemes. Many attempts to do so have assumed very simple target structures and have required broadband data over an impracticably large set of aspect angles [c.f. 2,3]. But in most missile applications, data are limited to high frequencies (wavelength much less than target size) and can only be collected from a small range of unknown target orientations. Moreover, real targets are usually very complex structures consisting of distributed scatterers with

*Research Department, Physics Division Code 3814, Naval Weapons Center, China Lake, CA 93555–6001. This work was supported (in part) by funding supplied by the Office of Naval Research.

varying reflectivities and associated intrinsic phase shifts. In fact, so little of the target's characteristics can be determined with certainty that the problem must be approach statistically.

Below, we discuss the glint problem in some detail and show how it can be related to the problem of target classification. Section 2 establishes our notation and formally develops a connection between large glint error and amplitude fades in the scattered field. In section 3 this approximate relationship between glint and amplitude is used to explain the validity of many of the currently proposed glint moderation schemes. Section 4 fashions an improved glint reduction scheme appropriate for "single frequency" data. The effectiveness in reducing direction angle error is demonstrated with simulated data in section 5. In sections 6 and 7 we examine the statistics of the phase derivative data and show how these statistics depend on target structure. This dependence will serve as the basis for a scheme of target classification based on the extremely limited radar data available to missile systems.

2. Scattered Fields and Glint. Since glint effects are often the limiting factor in tracking performance, a great deal of effort has been expended in trying to eliminate them. Typically, correction schemes have been *ad hoc* and have made use of general observed behavior of glint error:

- as a function of time, the glint error consists of a "smoothly" varying component punctuated by sharp glint "spikes"

- as a function of frequency, glint error displays the same spike-interrupted smoothness

- there is a strong correlation between these glint spikes and deep amplitude fades

Most of these correction schemes use the spike/amplitude correlation as an independent indicator of large glint error.

In the weak-scatterer approximation we can write the far-field scattered response $E(\mathbf{k})$ due to a harmonic excitation of a target as a linear superposition of waves with strength $|A(\mathbf{r})|$ radiating from locations $\mathbf{r}$:

$$(1) \qquad E(\mathbf{k}) = \int_D A(\mathbf{r}) \exp(i\mathbf{k} \bullet \mathbf{r}) d^3 r \ .$$

Here, D is the support of $A(\mathbf{r})$ and $\mathbf{k}$ is the wave vector with $|\mathbf{k}| = 2\pi/\lambda$. (We have suppressed the $\exp(-i\omega t)$ time dependence.) This Fourier transform relationship between the target function $A(\mathbf{r})$ and $E(\mathbf{k})$ can be used to continue the scattered field into complex n-dimensional space by setting $\mathbf{k}$ complex. When D is of finite extent the Plancherel–Pólya theorem [4] allows us to immediately classify this scattered field $E(\mathbf{k})$ as an entire function of exponential type.

For a target rotating about an axis orthogonal to the $x-y$ plane with aspect angle θ, equation (1) becomes

$$E(k,\theta) = \int_D A(x,y) \exp[ik(x \ \sin\theta - y \cos\theta)] dx dy \ ,$$

where now, $A(x, y)$ represents the source function integrated along the axial direction. If we denote $\xi = k \sin \theta$ and $\psi = -k \cos \theta$, then we can write this as the iterated integral

$$(2) \qquad E_\psi(\xi) = \int_a^b A_\psi(x) \exp(i\xi x) \, dx \; ,$$

where

$$A_\psi(x) \equiv \int A(x, y) \, \exp(i\psi y) \, dy \; ,$$

and each integral has limits appropriate to the target support. Since $A_\psi(x)$ vanishes for x outside of some finite region, then $E_\psi(\xi)$ is an entire function of exponential type in the variable ξ. Moreover, its logarithm is also analytic everywhere except at the branch points for which $|E_\psi(\xi)| = 0$ and, in particular, the derivative $\partial \ln E_\psi(\xi)/\partial \xi$ is analytic. Writing $E_\psi(\xi) = |E_\psi(\xi)| \exp[i\varphi(\xi)]$ yields

$$\frac{\partial \ln E_\psi(\xi)}{\partial \xi} = \frac{\partial \ln |E_\psi(\xi)|}{\partial \xi} + i \frac{\partial \varphi(\xi)}{\partial \xi} \; .$$

This function is also analytic everywhere except at the zeros of $|E_\psi(\xi)|$. If we integrate it over a path encompassing the upper half of the complex plane, we will need to take special care with its complex zeros. For finite ξ these zeros can be handled by modifying the contour [5] leaving only the zero at infinity ($E_\psi(z = re^{i\alpha} \to \infty) = 0$). However, since [6]

$$\lim_{r \to \infty} \frac{\ln |E_\psi(re^{i\alpha})|}{r} = 0 \qquad \text{and} \qquad \lim_{r \to \infty} \frac{\varphi(re^{i\alpha})}{r} = 0 \; ,$$

then, using L'Hopital's rule, the function vanishes at $r = \infty$ and by Titchmarsh's theorem [6] the real and imaginary parts of this derivative are a (modified) Hilbert transform pair. We have

$$(3) \qquad \frac{\partial \varphi(\xi)}{\partial \xi} = -\frac{1}{\pi} P \int_{-\infty}^{\infty} \frac{1}{(\xi' - \xi)} \frac{\partial \ln |E_\psi(\xi')|}{\partial \xi'} \, d\xi' + \frac{\partial \varphi(0)}{\partial \xi} + 2 \operatorname{Im} \sum R_j(\xi) \; ,$$

where $P \int (*)$ denotes the principal value integral, $R_j(z)$ is the functional contribution due to the j-th zero at z_j and the sum is over all such contributions from complex zeros in the upper half-plane. As $|z| \to \infty$, the zeros of $|E_\psi(z)|$ converge to the real axis [7] and consequently, for very large ξ, the integral term in (3) dominates the collective effects of the residues.

It is now straightforward to use this result to obtain an expression relating glint to scattered field amplitude. However, the relationship is *noncausal* and can be used to determine glint only with a knowledge of both past and future values of the amplitude. For our purposes, we need to replace (3) by a local approximation. To do this, we treat (3) phenomenologically.

Because the strongest contribution to the integral term in (3) comes from the deep fades in the field amplitude, an accounting for these fades is especially important as they are associated with large glint error. Because these fades are isolated, we can make the approximation

$$-\frac{1}{\pi} \, P \int_{-\infty}^{\infty} \frac{1}{(\xi' - \xi)} \frac{\partial \ln |E_\psi(\xi')|}{\partial \xi'} d\xi' \approx -\frac{1}{\pi} \left(\frac{\partial \ln |E_\psi(\xi)|}{\partial \xi^+} - \frac{\partial \ln |E_\psi(\xi)|}{\partial \xi^-} \right) ,$$

where $\partial/\partial \xi^\pm$ denote left and right derivatives. Clearly, the quality of this approximation is diminished far from a fade. If $\ln |E_\psi(\xi)|$ is sufficiently smooth so that its first derivative is continuous, then we may make the further approximation

$$\left(\frac{\partial \ln |E_\psi(\xi)|}{\partial \xi^+} - \frac{\partial \ln |E_\psi(\xi)|}{\partial \xi^-} \right) \approx \frac{\partial^2 \ln |E_\psi(\xi)|}{\partial \xi^2} \, \Delta \xi \ .$$

For small angles, $\xi \approx k\theta$ and $\psi \approx -k$, and our local approximation becomes, in the high frequency limit,

$$(4) \qquad\qquad \frac{\partial \varphi_k(\theta)}{\partial \theta} \quad \propto \quad \frac{\partial^2 \ln |E_k(\theta)|}{\partial \theta^2} \ .$$

But the derivative of phase with respect to aspect is the measure of target direction information that is to be used and is denoted by $g_k(\theta)$. For a radar directed at the target, therefore, we have related the glint $g_k(\theta)$ of the (high frequency) scattered field to its magnitude.

3. Frequency Diversity and Glint Reduction. The approximate result (4) should be quite accurate for those θ corresponding to glint spikes, and (performing the differentiation) we have

$$(5) \qquad g_k(\theta) \quad \propto \quad \frac{1}{|E_k(\theta)|} \frac{\partial^2 |E_k(\theta)|}{\partial \theta^2} - \left(\frac{1}{|E_k(\theta)|} \frac{\partial |E_k(\theta)|}{\partial \theta} \right)^2 \ .$$

This demonstrates the correlation between glint and amplitude fades: as the amplitude drops to zero, the magnitude of its derivative increases and the additional factor of $1/|E_k(\theta)|$ causes the glint to "spike" at the amplitude zero.

Since amplitude derivative information of this kind is completely lacking in radar tracking systems, we must content ourselves with using only the $1/|E_k(\theta)|$ factor in glint dependence. An easy argument [8] shows that $g_k(\theta)$ also depends in a sensitive way on the parameter k. This means that $g_k(\theta)$ acquired at fixed θ and for *multiple* frequencies, by virtue of the averaging property, should be expected to give better target direction information than single frequency glint. The improvement depends on the number of frequencies used, their spacing, and the method by which the g_k are averaged.

For a set of measurements $\{g_k : k = 1, \ldots, N\}$, we take the "best guess" g^{bg} as

$$g^{bg} = \sum_{k=1}^{N} w_k g_k ,$$

where w_k are the weights assigned to each frequency measurement g_k. Aside from simple averaging, for which $w_k = 1/N$, the obvious choice is to use weights based on the scattered field amplitude $|E_k(\theta)|$. Loomis and Graf [9] examined a variety of such weighting schemes. Of of the weights investigated, they discovered that "largest amplitude" weighting for which $w_k = \delta(|E_k| - |E_{\max}|)$ most-reduced the RMS tracking error (here, $|E_{\max}| \equiv \max\{|E_k| : k = 1, \ldots, N\}$). We have verified these results in simulation.

Unfortunately most missile tracking systems use too narrow bandwidth to allow for acquisition of enough g_k for the averaging to be of benefit. Therefore, a question of great importance concerns the application of these ideas to *"single frequency"* tracking systems.

4. "Aspect Diversity" Glint Reduction. At single frequencies, we must consider only aspect-diversity methods. From (5), we have the correspondence between the location of field zeros and glint spikes. In turn, a theorem due to Titchmarsh [10] describes the asymptotic density of these zeros. For the problem at hand, this result yields the number of zeros Δn located in the aspect interval $\Delta\theta$, in the high frequency limit, as

$$\Delta n \approx \frac{k(b-a)}{\pi} \, \Delta\theta \, ,$$

where $(b - a)$ is the target's crossrange extent (i.e., the support of the integrand in (2) is (a, b)). For a target maneuvering with instantaneous angular velocity $\Omega(t)$, we can write the approximate number of glint spikes in the time interval Δt as

$$(6) \qquad \Delta n \approx \frac{k(b-a)}{\pi} \, \Omega(t)\Delta t \, .$$

This target motion will also induce a doppler frequency shift. For scatterers located a (crossrange) distance x from the rotational axis, the shift in scattered field frequency obeys

$$\frac{\omega_D}{\omega_0} = \frac{2\Omega x}{c} + \quad \text{a term due to translational velocity} \, ,$$

where ω_D is the doppler frequency shift, ω_0 is the frequency of the interrogating signal, and c is the velocity of wave propagation. The difference in frequency shifts across the target is therefore given by

$$\frac{\Delta\omega_D}{\omega_0} = \frac{2\Omega(b-a)}{c} \, ,$$

and (6) can be written as

$$\Delta n \approx \frac{\Delta\omega_D}{2\pi} \, \Delta t \, .$$

If we wish to sample at a rate that guarantees that, out of N samples, not all N will correspond to a glint spike, then we should choose a sampling interval Δt as

$$(7) \qquad \Delta T = \frac{2\pi}{N\Delta\omega_D} \, .$$

A 10 meter target, for example, rotating at a rate of $2\pi/10$ radians per second about its center, would require a sampling interval of approximately $2 \times 10^{-3}/N$ seconds at X-band. This represents a fairly rapid aircraft maneuver and so we can expect that most sample intervals would be longer.

An aspect diversity scheme should now be clear: set the tracker pulse repetition frequency (PRF) to a rate sufficiently fast so that there is at least one pulse per time interval ΔT defined by (7). The "best" target location information will be the weighted average of N pulses, with one pulse selected from each of N consecutive intervals of duration ΔT. When the target is maneuvering, $\Delta\omega_D$ increases and, consequently, so does the frequency of the target location information. When the target is not maneuvering, $\Delta\omega_D$ decreasing - as does the target position data - but since the seeker is "on target" anyway, it doesn't matter. The key lies in relying only on the correctly *weighted* average for heading correction.

5. Simulation Results. Simulated scattering data were created by assigning relative scatterer strength and phase to collections of point scattering centers fixed to locations on computer generated targets. These targets were rotationally maneuvered, and the appropriate glint data was generated for the orientations associated with time steps ΔT incremented according to equation (7). Because these data were to be used for simple verification purposes, only one-dimensional data corresponding to linear targets rotating about fixed axes were considered.

By "target-pointing error" we mean the angular difference between actual and estimated target position. Our estimated pointing angle g^{bg} at time T_i is found by averaging over N successive glint measurements $g_k = g(T_i + k\Delta T_i), k = 1, 2, \ldots, N$:

$$g^{bg}(T_i) = \sum_{k=0}^{N-1} w_k g_k , \qquad (T_{i+1} \equiv T_i + N\Delta T_i).$$

We define the RMS pointing error, after M estimates, to be

$$\sigma^{bg} = \left[\frac{1}{M} \sum_{i=1}^{M} (g^{bg}(T_i) - \mathcal{A}(T_i))^2 \right]^{1/2} ,$$

where $\mathcal{A}(T_i)$ denotes the actual angular position of the target. Clearly, this error should depend on the number of measurements N used to determine g^{bg}. We denote the "ordinary" glint RMS error as $\sigma = \sigma^{bg}(N = 1)$.

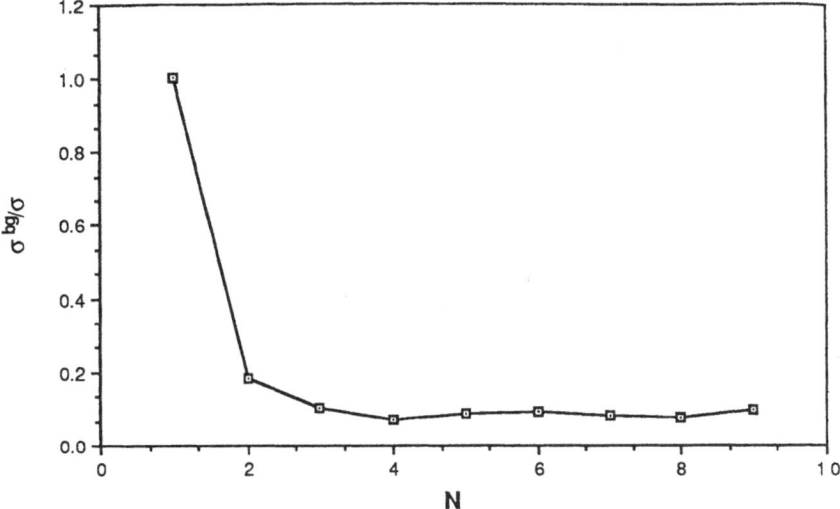

Figure 1: Relative glint error reduction as a function of sample number N. Simulated target was 10 meters long, consisted of 5 point scattering centers, and rotated at from 0 to 3 radians per second. Simulated frequency was $10GHz$. σ^{bg} was calculated using largest amplitude weighting.

Figure 1 displays the typical reduction in RMS pointing error that can be achieved under ideal (computer-simulated) conditions. For this figure, 1000 glint "measurements" were calculated corresponding to a 10 meter target made up of 5 scattering centers with random local strength and phase shift. σ^{bg} was calculated using the largest amplitude weighting scheme of Loomis and Graf, $w_k = \delta(|E_k| - |E_{\max}|)$. The synthesized interrogating signal frequency was 10 GHz, no attempt was made to simulate noise, and it was assumed that the doppler bandwidth $\Delta\omega_D$ could be determined with complete accuracy. The relative glint error reduction achieved by this scheme was essentially identical to that achieved in the Loomis and Graf frequency-agility method [9]. The ratio σ^{bg}/σ decreased monotonically to about $N = 4$, and then seemed to level-off. In some cases, it even worsened thereafter. Loomis and Graf apparently attribute this to large glint error in regions improperly sampled. A more likely explanation is that the amplitude weighting approximation is itself based on an approximation (5) which only accounts for the "spike-like" component of glint. It makes no effort to account for the lower frequency, smoothly varying, component of glint error and so cannot correct for it. Moreover, because this lower frequency component of glint cannot be locally estimated (it is due mostly to the residue term of equation (3)), it seems unlikely that σ^{bg}/σ can ever be made smaller than about .1 in practical systems.

6. Statistics of the phase derivatives. We now turn to the problem of automatic target classification using missile-based radar data. This radar data will usually be obtained at high frequencies and so will depend sensitively on target aspect. Because accurate aspect information is not available from noncooperating targets, the classification problem must be treated statistically.

A statistical target classification method based on the amplitude part of the radar return cannot be expected to be of much use in shape determination for complex targets because scattering amplitude depends in a delicate and complicated way on the structure and morphology of the local scattering center. However, it has long been realized that a wealth of target-shape information is contained in the phase of the asymptotic (high frequency) scattered field. The trouble is that absolute phase information is very difficult to obtain with any accuracy over a set of aspects. On the other hand, *phase derivative* information can be approximated to good accuracy by measuring the phase differences between closely spaced sensors. This is exactly the type of information that we have been examining above.

Let $E \equiv |E| \exp(i\varphi)$ denote the field scattered from the target. The phase is given by

$$\varphi = \tan^{-1}\left(-i\frac{E - E^*}{E + E^*}\right) ,$$

from which the phase derivatives can be easily determined once the aspect dependence has been specified. To do this, we again consider a plane wave originating from an observer and incident on a complex extended body made up of a collection of scattering centers. If $A_j \equiv a_j \exp(i\varphi_j)$, then in the case of a discrete scatterer equation (1) becomes

$$E(\mathbf{k}) = \sum_{j=1}^{N} a_j \, \exp[i(\mathbf{k} \bullet \mathbf{r}_j + \varphi_j)] .$$

In this equation, the target is taken to be composed of N scattering centers with associated scattering amplitudes a_j, positions $\mathbf{r}_j$ and intrinsic phase shifts φ_j.

If the observing radar lies at the position $\mathbf{R}$ from the target that lies at the origin of a standard Cartesian coordinate system then

$$\mathbf{b} \bullet \nabla_{\mathbf{R}} \varphi = \frac{su + tv}{s^2 + t^2} ,$$

where

$$s \equiv \sum_{j=1}^{N} a_j \cos[i(\mathbf{k} \bullet \mathbf{r}_j + \varphi_j)] , \qquad t \equiv \sum_{j=1}^{N} a_j \sin[i(\mathbf{k} \bullet \mathbf{r}_j + \varphi_j)] ,$$

$$u \equiv k \sum_{j=1}^{N} a_j \, \mathbf{b} \bullet \mathbf{r}_j \cos[i(\mathbf{k} \bullet \mathbf{r}_j + \varphi_j)] , \qquad v \equiv k \sum_{j=1}^{N} a_j \, \mathbf{b} \bullet \mathbf{r}_j \sin[i(\mathbf{k} \bullet \mathbf{r}_j + \varphi_j)] ,$$

and **b** is a unit vector orthogonal to **R**. The statistics of these phase derivatives can be determined by treating the $\{s, t, u, v\}$ as functions of the random variables $\{a_j\}$ and $\{\varphi_j\}$.

We take each of the random variables $\{a_j\}$ and $\{\varphi_j\}$ to be completely independent of the others. The $\{a_j\}$ are assumed to have the same statistical frequency distribution and the $\{\varphi_j\}$ are assumed to be uniformly distributed on $(-\pi, \pi)$. Under the stated assumptions, it is easy to show that $s, t, u,$ and v are statistically independent random variables with mean zero. Moreover, because each can be written as a sum $S = \Sigma \zeta_i$, where ζ_i are identically distributed with nonvanishing finite variances, then the central limit theorem allows us to conclude that, when the number of scattering centers is large, the $s, t, u,$ and v have normal frequency distributions. A straightforward calculation then shows that $\chi = \mathbf{b} \bullet \nabla \varphi$ obeys a student's t distribution law with probability frequency distribution

$$(8) \qquad f(\chi) = 1/2\Gamma_1^2(\Gamma_1^2 + \chi^2)^{-3/2} \ ,$$

where

$$\Gamma^2(\mathbf{b}) = k^2 \int_{-\infty}^{\infty} (\mathbf{b} \bullet \mathbf{r})^2 \rho(\mathbf{r}) d^3 r \ ,$$

and $\rho(\mathbf{r})$ is the scattering center density function weighted by local scatterer strength. (The details of these calculations can be found in [11,12].)

A word of explanation about the original assumptions is in order. For the random variables $\{a_j\}$ we have chosen noncommittal statistics. Furthermore, the independence of these variables is certainly reasonable on physical grounds. However, the uniform randomness of the $\{\varphi_j\}$ is somewhat arguable, and has been done to simplify the necessary mathematics. While *not incorrect* for a large class of realistic target models, the reader should observe that the assumption is equivalent to shifting each of the scattering centers along the (down-range) axis by some uniform random fraction of a wavelength. In the high frequency limit, then, this approximation can have few unhappy consequences.

7. Target-shape reduction. The statistics of the phase derivatives are completely determined by the parameter $\Gamma^2(\mathbf{b})$ which, according to (8), is proportional to a second moment of the weighted scattering center density function $\rho(\mathbf{r})$. In turn, this parameter Γ can be determined by applying maximum-likelihood estimation on a set of phase-derivative samples $\{\chi_i\}$ whose statistics are known to obey (8). This technique seeks the parameter Γ^2 that maximizes the product

$$\prod_{i=1}^{M} f(\chi_i) \ ,$$

where the index i ranges over the total number M of phase-derivative samples [11,12]. In practice we have found that relatively small data sets consisting of less than 50 data points are required to give a good estimate of Γ.

From equation (8) we see that as **b** varies, the graph of $\Gamma^2(\mathbf{b})$ will trace out an ellipsoid. (This is similar to the inertia ellipsoid.) Consider, for illustration purposes, a box-shaped target with length, width, and height equal to $2l, 2w$, and $2h$, respectively, and assume the scattering centers to be uniformly distributed throughout its volume. The characterizing ellipsoid of this "cloud" target is easy to compute, and we have

$$\Gamma^2(b) = \frac{k^2}{3R^2} \left(l^2 b_1^2 + w^2 b_2^2 + h^2 b_3^2 \right).$$

where the $b_i (i = 1, 2, 3,)$ are components of $\mathbf{b}$.

This ellipsoid has semi-axes proportional to the dimensions of the target and an orientation that is the same as that of the target. Note, however, that the ellipsoid cannot be uniquely determined from measurements for which $\mathbf{b}$ lies only in a plane. For aspect-limited data, we will obtain only the projection of the ellipsoid onto a plane. However, this limitation is rectifiable when there is only a little extra frequency information. The derivative of the phase of the scattered signal with respect to wave number obeys the same student's t statistics as the phase gradient [12]. Therefore, with data at only several frequencies and collected from aspects within a very *narrow* range, we can statistically determine the "inertia" ellipsoid representing the target. From this, the target may be categorized within the class of all targets having the same ellipsoid.

8. Conclusions. In the foregoing analysis we have considered the problem of glint error reduction in narrowband (single frequency) tracking systems. Narrowband systems cannot make use of frequency diversity methods which typically require more information than narrowband systems are physically able to acquire. However, we have shown how techniques devised for multiple-frequency data sets can be applied to multiple-aspect data sets. A consequence of this investigation is the recognition that a knowledge of the relative rotation rate between target and seeker (which is responsible for glint) can be used to set the sampling interval that minimizes glint.

We have also devised a method for characterizing the shape of a radar target based on information contained in the derivatives of its high-frequency echo phasefront return. Significantly, the technique works *best* when the target is complex (having many scattering centers), in contrast with other radar inverse scattering techniques [3] that work only with simple targets. Hence, the technique applies to a large class of practical situations. In addition, the method works even when the data are limited: only a small set of these phase derivatives obtained over a very narrow bandwidth are needed. Their aspect dependence is assumed to be unknowable and the only tacit requirement is that their aspect domain not be too large. (This is an advantage and not a limitation.)

The final objective of future research is an investigation of the degree to which target shape information is encoded in the high frequency scattered field. It is known, for example, that one-dimensional fields are completely determined by the location of their zeros. We have used these zeros, and the fact that at high frequencies the density of zeros become uniform, to help in moderating the glint. But the uniformity of the high-frequency zeros implies that no additional target shape information beyond that described above is available in the one-dimensional problem. In higher dimensions there are, in addition to the zero locations, also "form" factors [13]. A question of some importance concerns what additional structure information may be encoded in the asymptotic form of these factors (if any) and how it might be captured.

REFERENCES

[1] D.R. RHODES, *Introduction to Monopulse*, New York: McGraw-Hill, 1959.

[2] H.P. BALTES, *Inverse Scattering Problems in Optics, Topics in Current Physics*, H.P. Baltes, Ed. New York: Springer-Verlag, 1980.

[3] W-M. BOERNER, *Inverse Methods in Electromagnetic Imaging*, NATO ASI Series, Series C: Math. Phys. Sci., 143, Dordrecht, Holland: Reidel, 1985.

[4] M. PLANCHEREL and G. POLYA, *Fonctions Entières et Intégrals de Fourier Multiples*, Comment. Math. Helv., 9 (1936–37), pp. 224–248; 10 (1937–38), pp. 110–163.

[5] G. ARFKEN, *Mathematical Methods for Physicists*, third edition, Academic Press, New York, 1985.

[6] R.E. BURGE, M.A. FIDDY, A.H. GREENAWAY and G. ROSS, *The Phase Problem*, Proc. R. Soc. Lond. A, 350 (1976), pp. 191–212.

[7] F.E. BOND and C.R. CAHN, *On Sampling the Zeros of Bandwith Limited Signals*, IRE Trans. Inf. Theory, 4 (1958), pp. 110–113.

[8] B. BORDEN, *Diversity Methods in Phase Monopulse Tracking - A New Approach*, IEEE–AES, to appear.

[9] J.M. LOOMIS III and E.R. GRAF, *Frequency-Agility Processing to Reduce Radar Glint Pointing Error*, IEEE Trans. Aero. and Elec. Sys., 10 (1974), pp. 811–820.

[10] E.C. TITCHMARSH, *The Zeros of Certain Integer Functions*, Proc. Lond. Math. Soc., 25 (1925), pp. 283–302.

[11] B. BORDEN, *High-Frequency Statistical Classification of Complex Targets Using Severely Aspect-Limited Data*, IEEE Trans. Antennas and Propag., AP-34 (1986), pp. 1455–1459.

[12] B. BORDEN, *Statistically Based Inverse Scattering From Complex Symmetric Targets Using Severely Aspect-Limited High-Frequency Phase Derivative Data*, Inverse Problems, 3 (1987), pp. 623–631.

[13] W.F. OSGOOD, *Lehrbuch der Funktionentheorie*, Teubner, Liepzig, 1929.

IMAGING OF MEDIA THAT DIFFUSE AND
SCATTER RADIATION†

F. ALBERTO GRÜNBAUM*, PHILIP KOHN**,
J.R. SINGER# AND JORGE P. ZUBELLI‡

1. Introduction. This article is concerned with a procedure to determine the internal structure of objects that diffuse and scatter radiation, which was introduced in [1]. Let us recall that a crucial assumption in X-ray tomography is that the path of the probing particles is rectilinear. This is due to the fact that the X-ray photons have very high energy. On the other hand, for many applications it would be desirable to use photons that have less energy. This gives a situation where diffusion and scattering are important features of the propagation inside the object. Indeed, these features may have important new diagnostic values for a clinician. The presence of a significant amount of diffusion and scattering requires the consideration of infinitely many nonrectilinear paths joining the source to the detector. We consider here a simple probabilistic model, which admittedly could be an oversimplification. More realistic models are being considered by some of us. These models could be applicable to a fairly broad range of physical situations. We stress that the model presented here is substantially different from the one used in traditional X-ray tomography. The corresponding inverse problem that we obtain is nonlinear in contradistinction with the one in X-ray tomography. For information and literature on the latter we refer the reader to [2,3,4,5,6,7,8] and references therein. The model we present here is an attempt to interpret some experiments that were done by some of us with the propagation of infrared light through tissues. It turns out that at certain frequencies one could see, for thin tissues, shadows of the internal structure. As the thickness increased those became fainter and traditional reconstruction techniques did not work. The model proposed below considers each possible path from an outside source to an outside detector and associates to it a probability. This probability depends on the scattering and attenuation characteristics of the points along this path. However, instead of considering the object as a continuum, we discretize it in the spatial variables as well as in the possible directions of movement of the particles. A more ambitious approach might use a continuous model. But even then we expect that ultimately the resulting equations would have to be treated numerically and discretization would be unavoidable.

The plan for this article is the following: In Section 2 we describe our model in detail. In Section 3 we describe the solution of the forward problem associated with this model, i.e. the problem of finding the radiation intensity measured at a detector

†Expanded version of a talk delivered by J.P. Zubelli at the IMA Summer Program on Radar and Sonar: June 18–29, 1990.
*Dept. of Mathematics, University of California at Berkeley, CA 94720, USA.
**I.C.S.I., University of California at Berkeley, CA 94720, USA.
#E.E.C.S. & Biophysics Depts., University of California at Berkeley, CA 94720, USA.
‡Dept. of Mathematics, University of California at Santa Cruz, CA 95064, USA.

from a unit source given the absorption and scattering characteristics of the medium. In Section 4 we discuss one possible approach to the inverse problem. Section 5 is concerned with examples of numerical reconstructions. Before embarking on a more detailed description of the model, we point to a somewhat related effort in [9].

2. The Model. In this section we present a two-dimensional version of the model. However, it immediately carries over to the more realistic three-dimensional situation. It goes as follows:

1. Divide the object into planar elements, which we call pixels (or voxels in the three-dimensional case). Typically, the size of the pixels equals the desired resolution. For definiteness we take $x_i = i\rho$, $y_j = j\rho$, and

$$\Delta_{ij} = \left\{ (x,y) \in \mathbf{R}^2 | x_i - \frac{1}{2}\rho < x < x_i + \frac{1}{2}\rho \text{ and } y_j - \frac{1}{2}\rho < y < y_j + \frac{1}{2}\rho \right\},$$

where ρ is the pixel size.

2. Assume that the radiation consists of a large number of particles that enter the object at an arbitrary source position. Once it exits the object, if not caught by a detector a particle does not re-enter the object.

3. A particle moves by going from one pixel to an adjacent pixel within each pixel the particle may be annihilated by absorption. The probability of not being absorbed in pixel Δ_{ij} will be called ω_{ij}. See Figure 2.1.

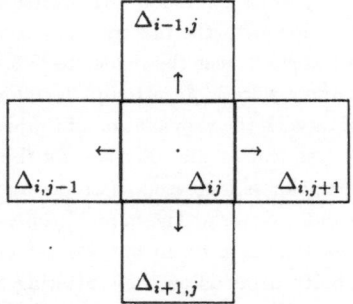

Figure 2.1. Movement of a single particle.

4. Once a particle enters Δ_{ij} and is not absorbed, it has a probability $\left\{ \begin{matrix} f_{ij} \\ b_{ij} \\ s_{ij} \end{matrix} \right\}$ of going $\left\{ \begin{matrix} \text{forward} \\ \text{backwards} \\ \text{sideways} \end{matrix} \right\}$. See Figure 2.2.

We note that in two dimensions

(1) $$b_{ij} + 2s_{ij} + f_{ij} = 1,$$

whereas in three dimensions

(2)
$$b_{ijk} + 4s_{ijk} + f_{ijk} = 1.$$

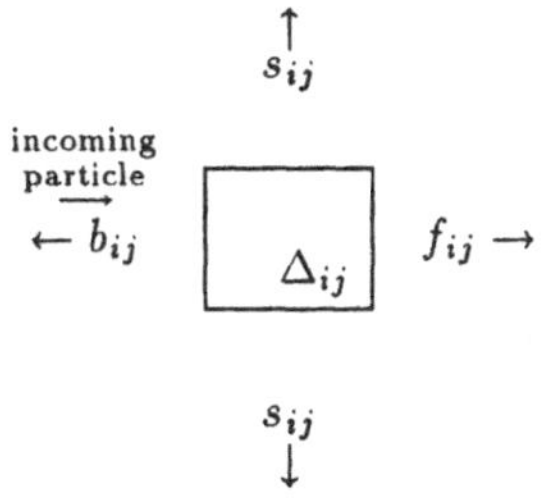

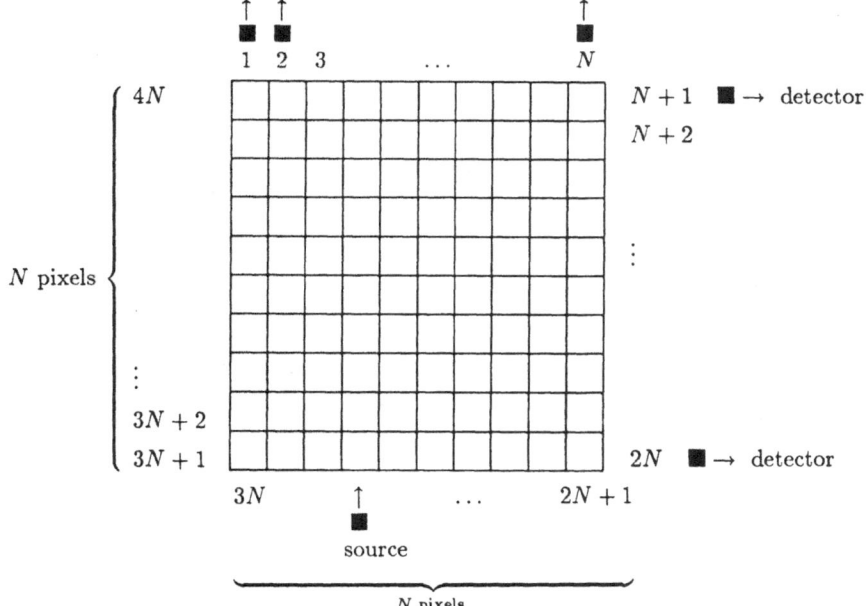

Figure 2.2. Different probabilities associated to the movements of a particle.

5. We can measure, for all possible source-detector pairs the relative intensity of the transmitted radiation. This intensity is assumed to equal the probability of going from the source to the detector in consideration.

Figure 2.3. Diagram of sources and detectors in generic positions, with the convention for their numbers.

From the aforementioned model, the object is characterized by a set of $3N^2$ variables, $(\omega_{ij}, f_{ij}, s_{ij})_{1 \leq i,j \leq N}$. There are $16N^2$ possible positions for source-detector pairs. We shall see later that in order to perform a reconstruction not all of such $16N^2$ measurements are necessary.

The quantities $\omega_{ij}, f_{ij}, s_{ij}$ and b_{ij} are subject to certain inequalities due to their probabilistic nature:

$$(3) \qquad\qquad 0 \leq \omega_{ij}, f_{ij}, b_{ij}, s_{ij} \leq 1.$$

If $\omega_{ij}, f_{ij}, s_{ij}$ and b_{ij} satisfy (3) and also equation (1) (or (2) in the three-dimensional case), we shall say that $(\omega_{ij}, f_{ij}, s_{ij}, b_{ij})$ belongs to the *physical region*. It is important to remark that the case $\omega_{ij} = 1$ is critical because this would mean no attenuation at all. We also notice that if $s_{ij} = b_{ij} = 0$, for all i, j, we get the standard X-ray tomographic model.

3. The Direct Problem. The direct problem is: Given a distribution of parameters $(\omega_{ij}, f_{ij}, s_{ij})_{1 \leq i,j \leq N}$, find the probability that a particle injected at source k will be captured by detector ℓ. The solution of this problem is obtained by introducing "internal variables" and setting up a system of linear equations. These internal variables are defined as follows: $z_{ij}^{m,\ell}$ denotes the probability of being captured by a single detector placed at the ℓ-th position when entering the pixel Δ_{ij} from the direction m. For definiteness we employ the numbering of the detectors (and sources) as shown in Figure 2.3, and we take the directions $m = 1, 2, 3, 4$ according to the convention of Figure 3.1.

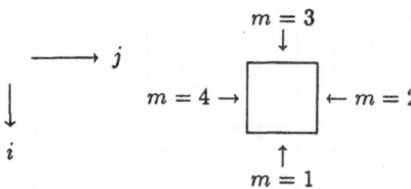

Figure 3.1. Ordering convention for entering a pixel.

If we consider a pixel situated in the interior of the object, it is not hard to write down its *balance equation* according to our model:

$$(4) \qquad \begin{bmatrix} z_{i,j}^{1,\ell} \\ z_{i,j}^{2,\ell} \\ z_{i,j}^{3,\ell} \\ z_{i,j}^{4,\ell} \end{bmatrix} = \omega_{ij} \begin{bmatrix} f_{ij} & s_{ij} & b_{ij} & s_{ij} \\ s_{ij} & f_{ij} & s_{ij} & b_{ij} \\ b_{ij} & s_{ij} & f_{ij} & s_{ij} \\ s_{ij} & b_{ij} & s_{ij} & f_{ij} \end{bmatrix} \begin{bmatrix} z_{i-1,j}^{1,\ell} \\ z_{i,j-1}^{2,\ell} \\ z_{i+1,j}^{3,\ell} \\ z_{i,j+1}^{4,\ell} \end{bmatrix}$$

For pixels on the periphery of the object, we must introduce boundary conditions. The equations look formally like (4), except that some of the values $z_{r,s}^{m,\ell}$ on the right-side of (4) are taken to be zero or one, according to the presence of the ℓ-th detector near to Δ_{ij}. We take $\bar{z}^\ell$ to be the vector whose components are the numbers $z_{i,j}^{m,\ell}$ with $1 \le i,j \le N$ and $1 \le m \le 4$. The problem of finding the probabilities $p_{\ell k}$, with $1 \le \ell, k \le 4N$, reduces to solving $4N$ systems of equations of the form

$$(5) \qquad \qquad A\bar{z}^\ell = \bar{v}^\ell,$$

where A depends on $(\omega_{ij}, f_{ij}, s_{ij})_{1\le i,j\le N}$ and $\bar{v}^\ell$ depends on the position of the detector and adjacent values of ω, f and s.

Of course, the first natural question is whether or not the matrix A is invertible. The answer to this question is given in terms of the following:

LEMMA. If $(\omega_{ij}, f_{ij}, s_{ij}, b_{ij})$ belongs to the physical region and $\omega_{ij} < 1$, for $1 \le i,j \le N$, then A is invertible and the solutions of (5) depend analytically on the parameters $(\omega_{ij}, f_{ij}, s_{ij})_{1\le i,j\le N}$.

The proof of this lemma is trivial because if $|\omega_{ij}| < 1$, for all $1 \le i,j \le N$, then the matrix A is strictly diagonally-dominant, with ones along the main diagonal. The analytic dependence of the solutions on the variables $(\omega_{ij}, f_{ij}, s_{ij})_{1\le i,j\le N}$ follows from the algebraic dependence of the matrix A on these variables.

4. The Inverse Problem. We can state the inverse problem for the model presented in Section 2 as follows:

Given as data the measurements of the $16N^2$ possible pairs of sources and detectors, or a fraction of such measurements, recover the $3N^2$ quantities characterizing the object.

This problem is highly nonlinear and even in situations like $N = 2$ far from trivial. Some explicit formulae were given in [10,11].

The first important question is: How redundant is the data given by $(p_{\ell k})_{1\le\ell,k\le 4N}$? A partial answer is given by the following[1]:

PROPOSITION. Given a distribution $(\omega_{ij}, f_{ij}, s_{ij}, b_{ij})_{1\le i,j\le N}$ in the physical region, with $\omega_{ij} < 1$, the measurement obtained by placing a detector at position ℓ and a source at position k equals the one obtained by reversing their positions. In other words $p_{\ell k} = p_{k\ell}$.

Proof. The proof is a direct consequence of the probabilistic interpretation of $p_{\ell k}$. Namely, let $\alpha = ((i_k, j_k), \ldots, (i_\ell, j_\ell))$ be a path joining the k-th source to the ℓ-th detector, as shown for example in Figure 4.1.

[1]The authors would like to acknowledge very helpful conversations with Geoff Latham and Harold Naparst on this issue.

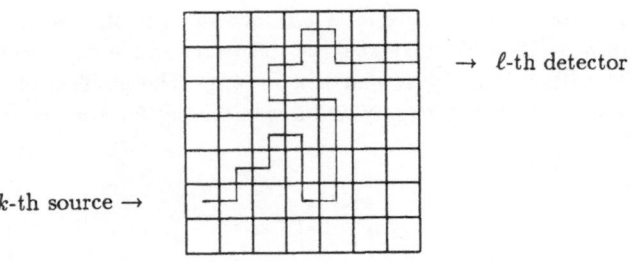

k-th source $\rightarrow$

$\rightarrow$ ℓ-th detector

Figure 4.1. One path joining a source to a detector.

Given a pixel (i,j) in α, we denote by t_{ij} the value of the different transition probabilities f_{ij}, s_{ij}, and b_{ij} according to the movement of the particle along the path to the subsequent site. The probability that the particle will emerge at the ℓ-th detector by travelling on α is given by the product of all values $\omega_{ij}t_{ij}$ with (i,j) in α. By (i,j) in α, we mean (i,j) one of the entries in the ordered n-tuple α. If we denote by $\mathcal{A}_{\ell k}$ the set of all paths joining ℓ to k we have that

$$p_{\ell k} = \sum_{\alpha \in \mathcal{A}_{\ell k}} \prod_{(i,j) \in \alpha} \omega_{ij}t_{ij}$$

The crucial point now is that the probability of turning right equals the probability of turning left. Therefore, we can say that for each path in $\mathcal{A}_{\ell k}$, the reverse path which is in $\mathcal{A}_{k\ell}$ has exactly the same probability. Hence, $p_{k\ell} = p_{\ell k}$. $\square$

The proposition implies that we can restrict our consideration to only $8N^2 + 2N$ of the $16N^2$ possible measurements. Still, because we are trying to determine only $3N^2$ variables, we have an overdetermined problem. This overdetermination works more as a blessing than as a curse, because in practice one expects to find a good amount of noise and inaccuracy in the measurements. This should be improved by redundancy. The three-dimensional case presents an even more dramatic relation between the amount of data available for the reconstruction and the number of unknowns. Indeed, in this case, for a cube of N voxels per edge, we have $18N^4 + 3N^2$ pieces of data[2] to reconstruct a total of $3N^3$ unknowns.

Our approach to the inverse problem consists in using a nonlinear least-squares minimization technique. In broad terms, let $(\omega_{ij}, f_{ij}, s_{ij})_{1 \leq i,j \leq N}$ be the true values of the characteristics of the object. We start from some initial guess $(\omega_{ij}^o, f_{ij}^o, s_{ij}^o)_{1 \leq i,j \leq N}$, and obtain a sequence of iterates $(\omega_{ij}^n, f_{ij}^n, s_{ij}^n)_{1 \leq i,j \leq N}$, for $n = 1, 2, \ldots$. We stop when the error

(6)
$$E = \sum_{\ell=1}^{N_d} \sum_{k=1}^{\ell} (p_{\ell k} - c_{\ell k}^n)^2$$

is below a certain threshold, where $c_{\ell k}^n$ is the computed value of the measurements for the configuration $(\omega_{ij}^n, f_{ij}^n, s_{ij}^n)$. The movement is done according to the gradient.

---

[2] Taking into account the symmetry resulting from the straightforward extension of Lemma 2 to 3 dimensions.

In the practical examples, we used the Levenberg-Marquardt algorithm [12,13]. However, a number of other minimization techniques are available.

We conclude this section by proposing an algorithm to compute the Jacobian of the function

$$F : \bar{x} = (\omega_{ij}, f_{ij}, s_{ij})_{1 \leq i,j \leq N} \mapsto (p_{\ell k})_{1 \leq k,\ell \leq 4N}.$$

This Jacobian is important in a number of numerical methods used to minimize E. Obviously, explicit formulae for such Jacobians would be extremely complicated and useless.

To compute the Jacobian, one would in principle resort to the use of a finite difference approach to approximate the derivatives. Our goal is to propose an alternative method using the particularities of the problem at hand. Before we proceed, we calculate the complexity involved in computing the Jacobian of F by means of a differencing scheme. Recall that the cost of computing the L-U decomposition of a full-rank $n \times n$-matrix is roughly $\frac{n^3}{3}$ arithmetic operations. Once the L-U form has been found, the forward-elimination and back-substitution algorithms take about $\frac{n^2}{2}$ arithmetic operations each. To compute $F(\bar{x})$ we must solve equation (5) for $\ell = 1, \ldots, 4N$, and then pick the values of $p_{\ell k}$ from $\bar{z}^\ell$, so the computational cost involved in each such step is given by

$$\frac{(4N^2)^3}{3} + (4N - 1)(4N^2)^2 \cong \frac{64}{3} N^6.$$

Note that in this formula we are assuming that we perform the L-U decomposition of A the first time we solve the system, and from that point on we only use forward-elimination and back-substitution. Suppose now that one decides to compute the Jacobian by the method of finite differences. This amounts to repeating this process described above for each of the $3N^2$ arguments of F, incrementing one at a time and computing the Newton quotient

$$\frac{\Delta F}{\Delta x_i} = \frac{1}{\varepsilon} \{ F(x_1, \ldots, x_i + \varepsilon, \ldots, x_{3N^2}) - F(x_1, \ldots, x_{3N^2}) \}.$$

The complexity involved in this process is roughly

$$(3N^2 + 1) \left(\frac{64}{3} N^6 + 64N^5 - 4N^4 \right) \simeq 64N^8.$$

The algorithm we are about to describe relies on the following remarks:

i) Solving the inverse problem for $(\omega_{ij}, f_{ij}, s_{ij})_{1 \leq i,j \leq N}$ is equivalent to solving for the products $\bar{f}_{ij} \overset{\text{def}}{=} \omega_{ij} f_{ij}$, $\bar{s}_{ij} = \omega_{ij} s_{ij}$, and $\bar{b}_{ij} = \omega_{ij} b_{ij}$.

ii) If we differentiate (5) with respect to say $\bar{f}_{ij}$ we get:

$$A \frac{\partial \bar{z}^\ell}{\partial \bar{f}_{ij}} + \frac{\partial A}{\partial \bar{f}_{ij}} \bar{z}^\ell = \frac{\partial \bar{v}^\ell}{\partial \bar{f}_{ij}}$$

and hence

(7)
$$\mathsf{A}\frac{\partial \bar{z}^\ell}{\partial \bar{f}_{ij}} = \frac{\partial \bar{v}^\ell}{\partial \bar{f}_{ij}} - \frac{\partial \mathsf{A}}{\partial \bar{f}_{ij}} \bar{z}^\ell.$$

Similar equations hold for $\frac{\partial \bar{z}^\ell}{\partial \bar{s}_{ij}}$ and $\frac{\partial \bar{z}^\ell}{\partial \bar{b}_{ij}}$. It turns out that the dependence of A on $\bar{f}_{ij}$ is only as a linear factor which appears in a 4 by 4 minor, as can be seen from eq. (4). If the pixel Δ_{ij} is not on the boundary the terms $\frac{\partial \bar{v}^\ell}{\partial \bar{f}_{ij}}$, $\frac{\partial \bar{v}^\ell}{\partial \bar{s}_{ij}}$ and $\frac{\partial \bar{v}^\ell}{\partial \bar{b}_{ij}}$ vanish. Otherwise, at most four entries of $\bar{v}^\ell$ depend on such variables. In any case, the construction of the right-side of equation (7) is straightforward and the computations are negligible.

Our proposal for computing $\frac{\partial \bar{z}^\ell}{\partial \bar{f}_{ij}}$ is to use the L-U decomposition of the matrix A (which has to be computed anyway) and solve for $\frac{\partial \bar{z}^\ell}{\partial \bar{f}_{ij}}$ by means of equation (7) with the help of forward-elimination and back-substitution. The number of arithmetic operations to compute $\left(\frac{\partial \bar{z}^\ell}{\partial \bar{f}_{ij}}, \frac{\partial \bar{z}^\ell}{\partial \bar{b}_{ij}}, \frac{\partial \bar{z}^\ell}{\partial \bar{s}_{ij}} \right)_{1 \le \ell \le 4N}$ as well as $\bar{z}^\ell$ is roughly:

$$\frac{(4N^2)^3}{3} + (4N - 1)(4N^2)^2 + 4N \cdot 3N^2 \cdot (4N^2)^2 \cong 3 \times 64N^7.$$

Therefore, for large values of N the use of the method described provides a significant improvement. Similar considerations also hold for the three-dimensional situation.

5. A Few Numerical Examples. In this section we shall present a few examples of reconstructions. The main goal here is to study the numerical behaviour of the inverse problem and the noise sensitivity of the reconstruction.

The first example was done in two dimensions, and we solved for the attenuation, which is defined as $v_{ij} = 1 - \omega_{ij}$. See Figure 5.1 and Figure 5.2.

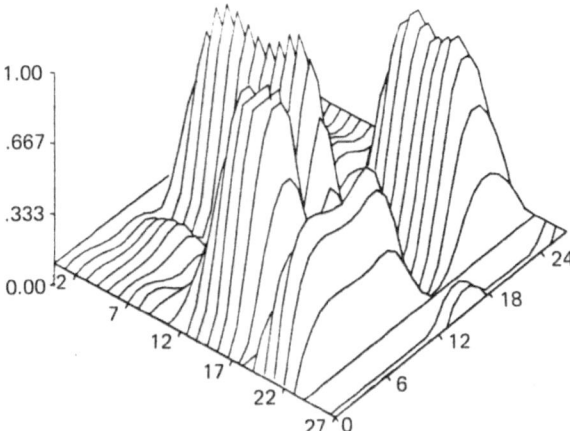

Figure 5.1. A simulated diffusing two-dimensional slice of an object (a phantom) that has attenuation values plotted as the hills, (heights). The x and y coordinates are those of the object.

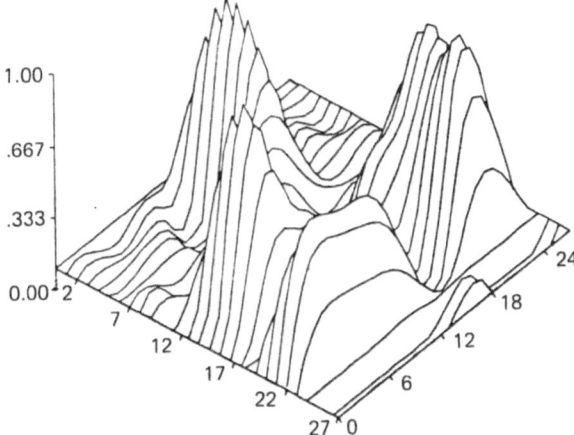

Figure 5.2. The reconstruction of the phantom object using the diffused radiation emergent from the phantom in Figure 5.1. The attenuation values are plotted as the hills. In this plot the correspondence can be compared with the phantom object of Figure 5.1.

To probe into the noise sensitivity of the inverse problem, we went through the following steps: Firstly, we chose a configuration $(\omega_{ij}, f_{ij}, s_{ij})_{1 \leq i,j \leq N}$ and computed the corresponding values of $(p_{\ell k})_{1 \leq k, \ell \leq N}$. Secondly, we introduced noise by multiplying the values of $(p_{\ell k})_{1 \leq \ell, k \leq N}$, by a random variable of the form $1 + \alpha X$, where X was a (pseudo) random number of zero mean and taking values in $[-1, 1]$. We took $\alpha = 0$ (i.e., no noise), 0.1×10^{-3}, 1×10^{-2}, and 5×10^{-2}. The initial guess in these simulations were constant configurations, i.e. , we did not use any a priori information. The results are displayed in Figures 5.3 through 5.11.

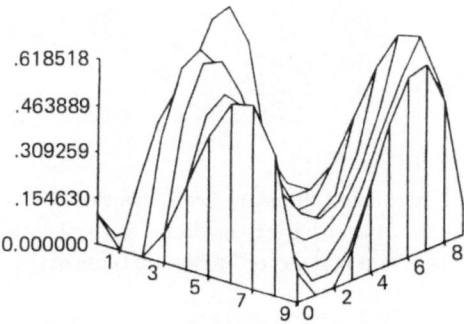

Figure 5.3. Characteristics of the attenuation of a cross section of a three dimensional object, (a phantom object).

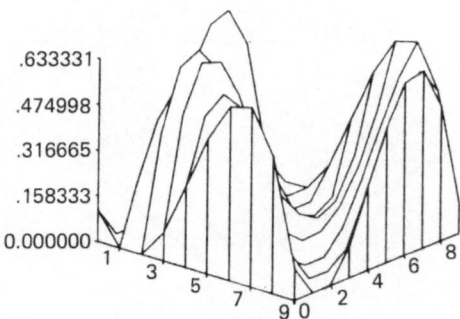

Figure 5.4. Reconstructed attenuation in the presence of 1% multiplicative noise.

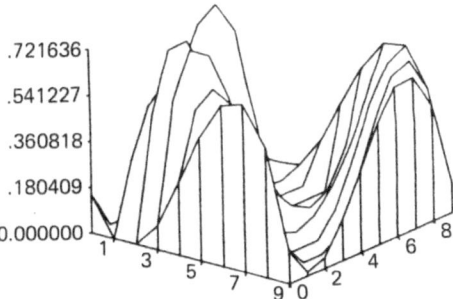

Figure 5.5. Reconstructed attenuation in the presence of 5% multiplative noise.

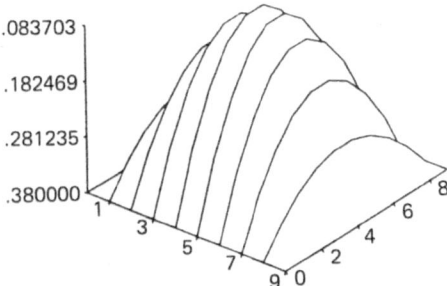

Figure 5.6. Characteristics of the forward-scattering coefficient of a cross section of a three-dimensional object (a phantom object).

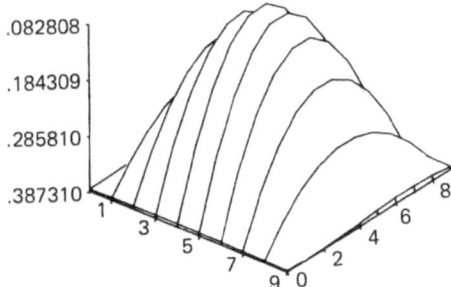

Figure 5.7. Reconstructed forward-scattering coefficient of a section of a three-dimensional object in the presence of 1% noise.

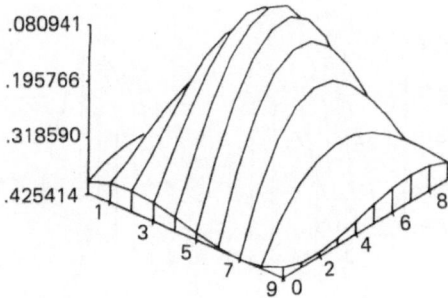

Figure 5.8. Reconstructed forward-scattering coefficient of a section of a three-dimensional object in the presence of 5% noise.

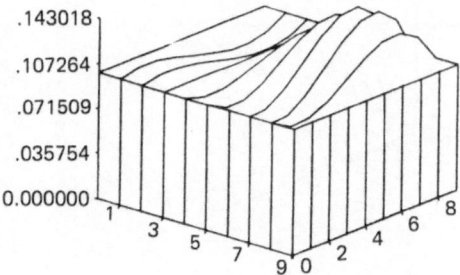

Figure 5.9. Characteristics of the side-scattering coefficient of a section of a three-dimensional object (a phantom object).

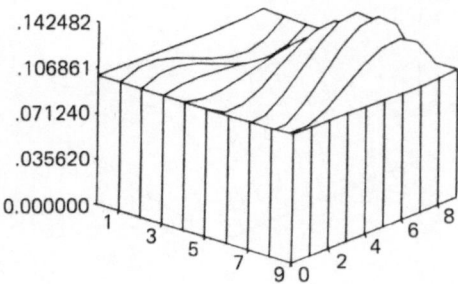

Figure 5.10. Reconstructed side-scattering coefficient of a section of a three-dimensional object in the presence of 1% multiplicative noise.

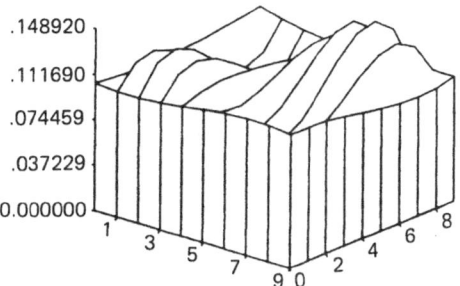

Figure 5.11. Reconstructed side-scattering coefficient of a section of a three-dimensional object in the presence of 5% multiplicative noise.

6. Conclusions and Final Remarks. It should be clear that the first theoretical question that needs to be addressed is the well-posedness of the inverse
problem. Based on an extensive number of numerical examples we have strong
reasons to believe that this problem is well-posed *provided* we are in fairly typical situations. We do not expect uniqueness, for example, to hold in situations
where there is a lot of symmetry. In fact, the configuration of Figure 6.1 provides
a counterexample in two-dimensions.

α	β	α
β	γ	β
α	β	α

Figure 6.1. A configuration that leads to nonunique solutions
of the inverse problem.

In this figure α and β denote pixels with common values for f_{ij}, b_{ij} and s_{ij}.
It is not difficult to see that the data obtained from this configuration leads to a
continuum of solutions to the inverse problem. We stress that this simple example contains a tremendous amount of symmetry, which is not expected in realistic
problems of reconstruction.

The next problem that must be overcome is the fact that the complexity of the
computations grows very fast with the number of pixels. In fact, the reconstructions presented here would not have been possible without the advent of very fast
computers. This complexity issue becomes particularly serious when we are faced
with the fact that the minimization of the error E involves an unspecified number
of iterations. Furthermore, the iterates may become trapped in local minima of
E, and yield misleading results. Consequently, it would be important to develop
algorithms for this specific problem that converge to the actual solution and avoid
local minima of E.

On the positive side, it is very important to note that the forward problem is very suitable for parallel computers. This is due to the fact that to compute its solution one can first perform the L.U. decomposition of the matrix A and then use forward-elimination and back-substitution for all the $4N$ different detector positions. This could be done in parallel, which could improve speed by as much as the order of $4N$ for the two-dimensional situation, and much better for the three-dimensional one.

Another positive point is that in many situations one has a certain amount of a priori information about the object, which could allow for a good initial guess: possibly a certain portion of the object is well known and one wants to reconstruct only a small number of variables. In these cases even the initial steps developed in the present article could be very helpful.

7. Acknowledgement. The authors would like to acknowledge the hospitality of the Institute for Mathematics and its Applications.

REFERENCES

[1] SINGER, J.R., GRÜNBAUM, F.A., KOHN, P., AND ZUBELLI, J.P., *Image Reconstruction of the Interior of Bodies that Diffuse Radiation*, Science, vol. 248 (1990), pp. 990–993.

[2] F. NATTERER, The Mathematics of Computerized Tomography, John Wiley and Sons (1986).

[3] CORMACK, A.M., *Representation of a function by its line integrals, with some radiological applications*, J. Appl. Phys., 34 (1963), pp. 2722–2727.

[4] HERMAN, G.T., Image Reconstruction from Projections. The Fundamentals of Computerized Tomography, Academic Press (1980).

[5] HOUNSFIELD, G.N., *Computerized transverse axial tomography: Part I, description of the system*, Br. J. Radiol, 46 (1973), pp. 1016–1022.

[6] LOGAN, B.F. AND SHEPP, L.A., *Optimal reconstruction of a function from its projections*, Duke Math. J., 42 (1975), pp. 645–659.

[7] SHEPP, L.A. AND LOGAN, B.F., *The Fourier reconstruction of a head section*, IEEE Trans. Nucl. Sci., NS-21 (1974), pp. 21–43.

[8] SHEPP, L.A. AND KRUSKAL, J.B., *Computerized Tomography: The new medical X-ray technology*, Am. Math. Monthly, 85 (1978), pp. 420–439.

[9] BARBOUR, R.L., GRABER, H. ARONSON, R. AND LUBOWSKY, J., *Model for 3-D optical imaging of tissue*, Proc. of International Geoscience and Remote Sensing Symposium, vol II (1990), pp. 1395–1399.

[10] GRÜNBAUM, F.A., *Tomography with diffusion*, Inverse Problems in Action, P. Sabatier editor, Springer-Verlag (1990), pp. 16–21.

[11] GRÜNBAUM, F.A., *Backscattering comes to the rescue*, Contemporary Mathematics, T. Quinto (ed.) to appear American Mathematical Society (1990).

[12] MORÉ, J.J., *The Levenberg-Marquardt Algorithm: Implementation and Theory*, Lecture Notes in Math. vol. 630, G.A. Watson, Ed., Springer-Verlag (1977).

[13] PRESS, W.H., FLANNERY, B.P., TEUKOLSKY, S.A.; VETTERLING, W.T., Numerical Recipes, The Art of Scientific Computing, Cambridge Univ. Press, New York (1987).

A NOTE ON GROUP CONTRACTIONS
AND RADAR AMBIGUITY FUNCTIONS

E.G. KALNINS† AND W. MILLER, JR.‡

Abstract. The Heisenberg group is a contraction of $G_A \times R$, where G_A is the affine group and R is the additive group of the real numbers. We work out the details of the contraction procedure and use it to show how the narrowband cross-ambiguity function arises as a limit of the wideband cross-ambiguity function. We also demonstrate how the expansion formula for the target distribution function in the narrowband case arises as a limit of the corresponding expansion formula for the wideband distribution function via the contraction procedure.

Key words. group contraction, radar ambiguity function

AMS(MOS) subject classifications. 22E70, 78A45, 43A70

1. Introduction. It is well known that the narrowband radar ambiguity and cross-ambiguity functions are approximations of the corresponding wideband functions. Some of the details of the approximations can be found in [1], [4], [20]. Here we shall employ the concept of group contraction to shed more light on the relationship between these functions. Further, we shall demonstrate how the expansion formula for the target distribution function in the narrowband case arises as a limit of the corresponding expansion formula for the wideband distribution function via the contraction procedure. Most of our computations will be formal, but the needed rigor can be supplied. We thank R. E. Blahut for suggesting this problem.

We review briefly the Doppler effect as it relates to radar and sonar. Consider a stationary transmitter/detector and a moving (point) target located at a distance $R(t)$ from the transmitter at time t. We assume that the distance from the transmitter to the target changes linearly with time,

$$(1.1) \qquad R(t) = r + vt.$$

Now suppose that the transmitter emits an electromagnetic pulse $s(t)$. The pulse is transmitted at a constant speed c in the ambient medium (e.g., air or water) impinges the target, and is reflected back to the transmitter/receiver where it is detected as the echo $e_{r,v}(t)$. (We assume $c > |v|$.) It is not difficult to show that the echo $e_{r,v}(t)$ takes the form

$$(1.2) \qquad e_{r,v}(t) = \sqrt{\frac{c-v}{c+v}} \, s\left(t\left[\frac{c-v}{c+v} \right] - \frac{2r}{c+v} \right),$$

†Department of Mathematics and Statistics, University of Waikato Hamilton, New Zealand

‡School of Mathematics, 127 Vincent Hall, University of Minnesota, Minneapolis, MN 55455, U.S.A. The work of this author was supported in part by the National Science Foundation under grant DMS 88–23054

where the scale factor $\sqrt{\frac{c-v}{c+v}}$ is needed if we require that the energy of the pulse is conserved:

$$\int_{-\infty}^{\infty} |s(t)|^2 dt = \int_{-\infty}^{\infty} |e(t)|^2 dt.$$

Changing notation, we write the echo as

$$(1.3) \qquad e_{x,y}(t) = \sqrt{y}s(y[t+x])$$

where $y = (c-v)/(c+v)$ and $x = -2r/(c-v)$. Note that the transformation $(x,y) \to (r,v)$ is one-to-one. The result (1.2) is the exact (wideband) solution for the given assumptions, [16], [17]. We can generalize this analysis to allow a distribution of moving targets with density function $D(x,y)$, where the support of this function is contained in the set $\{(x,y): y > 0\}$. Then the echo from the signal $s(t)$ is

$$(1.4) \qquad e(t) = \int_0^{\infty} \int_{-\infty}^{\infty} \sqrt{y}s(y[t+x])D(x,y)dx\,dy.$$

It is common to approximate the wideband solution under the assumption that $|v| \ll c$. (See [4] and [17] for careful analyses of the relationship between the wideband solution and the narrowband approximation.) One writes the signal in the form

$$s(t) = a(t)e^{2\pi i\omega_0 t}$$

where ω_0 is the mean frequency of the Fourier transform of s and $a(t)$ is assumed to be a slowly varying complex function of t (the *envelope* of the waveform) with respect to the exponential factor. In particular one requires that $a(y[t+x]) \approx a(t+x)$ over the (x,y) domain contributing to the integral (1.4).

Then we have

$$e_{x,y}(t) = \sqrt{y}a(y[t+x])e^{2\pi i(y[t+x])\omega_0}$$

or, since $y \approx 1 - 2\beta$, $\quad yx \approx -(2r/c)(1-\beta)$ where $\beta = v/c$, we have

$$e_{x,y}(t) \approx a(t - 2r/c)e^{2\pi i\omega_0[t(1-2\beta)-2(r/c)(1-\beta)]}$$
$$\approx s(t - 2r/c)e^{-4\pi i\beta\omega_0[t-2r/c]}.$$

This is called the *narrowband approximation* of the Doppler effect. (Generally speaking, the narrowband approximation is usually adequate for radar applications but is less often appropriate for sonar.) Similarly, we can derive an expression for the echo due to a distribution of moving targets in the narrowband approximation.

Suppose $\{s_n\}$ is a basis for $L_2(R)$, the standard space of square integrable functions on the real line. Then we can consider the cross-correlation of s_m with the echo e_n from s_n:

$$(1.5) \qquad A_{nm}(x,y) = \langle e_n, s_m \rangle = \sqrt{y}\int_{-\infty}^{\infty} s_n(y[t+x])\bar{s}_m(t)dt.$$

We call A_{nm} the *cross-ambiguity function* of s_n and s_m. Similarly the *(narrowband)* cross-ambiguity function is

(1.6)
$$\psi_{nm}(u,w) = \int_{-\infty}^{\infty} e^{2\pi itw} s_n(t+u)\, \bar{s}_m(t)\, dt$$

where $u = -2r/c$, $w = -2v\omega_0/c$ and we have ignored the phase factor $e^{8\pi ivr\omega_0/c^2}$. If $n = m$ these expressions are called *ambiguity functions*. The ambiguity and cross-ambiguity functions are of fundamental importance in the theory of radar and sonar. The use of the ambiguity function to estimate the range and velocity of point targets and, more generally, of the cross-ambiguity function to estimate target distribution functions in both the wideband and narrowband cases is basic to the theory, [3], [4], [14], [20].

There is an intimate relationship between the cross-ambiguity functions defined above and the theory of group representations. In particular the wideband cross-ambiguity functions can be interpreted as matrix elements of unitary irreducible representations of the two-dimensional affine group; the narrowband cross-ambiguity functions are matrix elements of unitary irreducible representations of the three-dimensional Heisenberg group. The basic properties of these functions thus emerge as consequences of analyses on affine and Heisenberg groups.

The affine group G_A is the matrix group with elements

$$A(a,b) = \begin{pmatrix} a & b \\ 0 & 1 \end{pmatrix}, \quad a > 0.$$

As is easy to check, the operators $\mathbf{L}[a,b]$ define a unitary representation of G_A on the Hilbert space $L_2(R)$:

(1.7)
$$(\mathbf{L}[a,b]\varphi)(t) = a^{-\frac{1}{2}}\varphi\left(\frac{t+b}{a}\right), \quad \varphi \in L_2(R).$$

The representation $\mathbf{L}$ is reducible, [9], [18]. Indeed it decomposes into a direct sum of two irreducible representations: $\mathbf{R}_+$ (whose representation space consists of those $\varphi \in L_2(R)$ whose Fourier transforms have support on the positive half line) and $\mathbf{R}_-$ (whose representation space consists of those $\varphi \in L_2(R)$ whose Fourier transforms have support on the negative half line). Note that the matrix element $\langle \mathbf{L}[a,b]s_n, s_m \rangle$ coincides with the wideband cross-ambiguity function (1.5) with $y = a^{-1}, x = b/a$.

The decomposition of the right regular representation of G_A is of interest for radar and sonar theory, [6], [7], [9], [10]. Here the representation space consists of functions $f(a,b)$ on G_A, square integrable with respect to the right invariant measure $da\, db/a$, $(0 < a < \infty, -\infty < b < \infty)$ and the group acts on this space by right multiplication. The decomposition can be written in the form

(1.8)
$$f(a,b) \equiv D(X,Y) = \sum_{n=0}^{\infty} (\langle \tilde{e}_n^-, \mathbf{L}[a,b]s_n^- \rangle + \langle \tilde{e}_n^+, \mathbf{L}[a,b]s_n^+ \rangle),$$

where $X = b$, $Y = a^{-1}$ and

$$\begin{aligned}
\tilde{e}_n(\tau) &= \int_{G_A} f(a', b')\mathbf{L}[a', b']\tilde{s}_n(\tau)d_\ell A \\
&= \int_{-\infty}^{\infty}\int_0^{\infty} f(a', b')\frac{1}{\sqrt{a'}}\tilde{s}_n\left(\frac{\tau + b'}{a'}\right)\frac{da'db'}{(a')^2} \\
&= \int_{-\infty}^{\infty}\int_0^{\infty} D(x, y)\sqrt{y}\tilde{s}_n(y[\tau + x])dy\,dx \\
&= F'(f)\tilde{s}_n(\tau).
\end{aligned}$$

(1.9)

Also, $\{s_n^+\}$ is an orthonormal basis for the subspace of $L_2(R)$ transforming according to $\mathbf{R}_+$ and $\{s_n^-\}$ is an orthonormal basis for the subspace transforming according to $\mathbf{R}_-$. Finally, $\tilde{S}_n^{\pm}(\omega) = \omega S_n^{\pm}(\omega)$ where $\tilde{S}_n^{\pm}$ and $S_n^{\pm}$ are the Fourier transforms of $\tilde{s}_n^{\pm}$ and $s_n^{\pm}$, respectively. If the $\{s_n\}$ belong to the Schwartz class then

$$\tilde{s}_n(t) = -\frac{i}{2\pi}\frac{ds_n}{dt}(t).$$

Note that (1.8) provides a scheme for determining the distribution $D(X, Y)$ experimentally: we can send out signals $\tilde{s}_n^{\pm}$, measure the echos $\tilde{e}_n^{\pm}$ and then cross-correlate these echos with the test signals $\mathbf{L}[a, b]s_n^{\pm}$ to reconstruct D.

It is not difficult to show [6], [10] that the operator $-\frac{i}{2\pi}\mathbf{L}^{-1}[a, b]F'(f)\frac{d}{dt}$ acting on Schwartz class functions is a Hilbert-Schmidt operator with finite trace. In particular, (1.8) is independent of the choice of orthonormal basis for $L_2(R)$, [5], [11], [12]. Choosing an arbitrary orthonormal basis $\{s_n\}$ in the Schwartz class, we obtain the more compact expansion

(1.10)
$$f(a, b) = \sum_{n=0}^{\infty}\langle\tilde{e}_n, \mathbf{L}[a, b]s_n\rangle.$$

The (three-dimensional real) Heisenberg group H_R is the group of matrices

$$B(x_1, x_2, x_3) = \begin{pmatrix} 1 & x_1 & x_3 \\ 0 & 1 & x_2 \\ 0 & 0 & 1 \end{pmatrix} \qquad x_j \in R.$$

For each real $\lambda \neq 0$, H_R admits a unitary irreducible representation by operators $\mathbf{T}^\lambda[x_1, x_2, x_3]$ acting on the Hilbert space $L_2(R)$:

(1.11)
$$\mathbf{T}^\lambda[x_1, x_2, x_3]\varphi(t) = e^{2\pi i\lambda(x_3 + tx_2)}\varphi(t + x_1), \quad \varphi \in L_2(R).$$

This is called the *Schrödinger representation* of the Heisenberg group, [9], [15].

It can be shown that

(1.12)
$$\int_{-\infty}^{\infty}\int_{-\infty}^{\infty} dx_1\,dx_2\langle\mathbf{T}^\lambda(x_1, x_2)\,\varphi_1, \varphi_2\rangle\langle\varphi_4, \mathbf{T}^\lambda(x_1, x_2)\,\varphi_3\rangle = \langle\varphi_1, \varphi_3\rangle\langle\varphi_4, \varphi_2\rangle$$

where

$$\langle \varphi_1, \varphi_2 \rangle = \int_{-\infty}^{\infty} \varphi_1(t) \, \overline{\varphi_2}(t) dt$$

is the usual inner product on $L_2(R)$ and $\mathbf{T}^\lambda(x_1, x_2) = \mathbf{T}^\lambda[x_1, x_2, 0]$, so that

$$(1.13) \qquad \langle \mathbf{T}^\lambda(x_1, x_2) \, \varphi_1, \varphi_2 \rangle = \int_{-\infty}^{\infty} e^{2\pi i \lambda t x_2} \varphi_1(t + x_1) \, \overline{\varphi_2}(t) dt.$$

If $\{s_n\}$ is an orthonormal basis for $L_2(R)$ then $\{\psi_{nm}(x_1, x_2) = \langle \mathbf{T}^\lambda(x_1, x_2)s_n, s_m \rangle \}$ is an orthonormal basis for $L_2(R^2)$, [2], [9], [14], [15]. It follows from (1.6) that narrowband cross-ambiguity functions are, essentially, matrix elements of the representation $\mathbf{T}^1$ of H_R.

If $f(x_1, x_2)$ belongs to $L_2(R^2)$ we can expand it in terms of the basis of cross-ambiguity functions to obtain

$$f(x_1, x_2) = \sum_{n,m} \int\int f(y_1, y_2) \langle \mathbf{T}^1(y_1, y_2)s_n, s_m \rangle dy_1 dy_2 \, \overline{\langle \mathbf{T}^1(x_1, x_2)s_n, s_m \rangle}$$

$$= \sum_n \langle \int\int f(y_1, y_2) \mathbf{T}^1(y_1, y_2)s_n \, dy_1 dy_2, \mathbf{T}^1(x_1, x_2)s_n \rangle$$

$$(1.14) \qquad = \sum_n \langle e_n, \mathbf{T}^1(x_1, x_2)s_n \rangle,$$

where

$$e_n(t) = \int_{-\infty}^{\infty} \int_{-\infty}^{\infty} f(y_1, y_2) \mathbf{T}^1(y_1, y_2)s_n(t) \, dy_1 dy_2$$

is the (narrowband) echo from the signal s_n due to the target distribution f.

We shall use group contractions to relate the expansion formulas (1.10) and (1.14).

2. The contraction procedure. We denote by $G_A \times R$ the direct product of the affine group G_A and the real line. A matrix realization of this group consists of the elements

$$(2.1) \qquad A(a, b, c) = \begin{pmatrix} a & b & 0 \\ 0 & 1 & 0 \\ 0 & 0 & e^c \end{pmatrix}, \qquad a > 0$$

with multiplication rule

$$A(a, b, c)A(a', b', c') = A(aa', ab' + b, c + c').$$

Here a, b and c are real numbers. A basis for the Lie algebra $L(G_A \times R)$ consists of the matrices

$$(2.2) \qquad A_1 = \begin{pmatrix} 1 & 0 & 0 \\ 0 & 0 & 0 \\ 0 & 0 & 0 \end{pmatrix}, \qquad A_2 = \begin{pmatrix} 0 & 1 & 0 \\ 0 & 0 & 0 \\ 0 & 0 & 0 \end{pmatrix}, \qquad A_3 = \begin{pmatrix} 0 & 0 & 0 \\ 0 & 0 & 0 \\ 0 & 0 & 1 \end{pmatrix}$$

with commutation relations

$$(2.3) \qquad [\mathcal{A}_1, \mathcal{A}_2] = \mathcal{A}_2, \quad [\mathcal{A}_1, \mathcal{A}_3] = \Theta, \quad [\mathcal{A}_2, \mathcal{A}_3] = \Theta$$

where Θ is the zero matrix, [8], [18]. We can pass from the Lie algebra to the group by means of the relation

$$(2.4) \qquad A(a, b, c) = \exp c\, \mathcal{A}_3 \exp b\, \mathcal{A}_2 \exp \ln a\, \mathcal{A}_1,$$

where

$$\exp c\, \mathcal{A} = \sum_{n=0}^{\infty} \frac{c^n}{n!} \mathcal{A}^n,$$

as the reader can verify.

The Heisenberg group H_R has the matrix realization

$$(2.5) \qquad B(x_1, x_2, x_3) = \begin{pmatrix} 1 & x_1 & x_3 \\ 0 & 1 & x_2 \\ 0 & 0 & 1 \end{pmatrix}, \quad x_j \in R$$

with group product

$$(2.6) \qquad B(x_1, x_2, x_3) B(y_1, y_2, y_3) = B(x_1 + y_1, x_2 + y_2, x_3 + y_3 + x_1 y_2).$$

A basis for the Lie algebra $L(H_R)$ consists of the matrices

$$(2.7) \qquad \mathcal{B}_1 = \begin{pmatrix} 0 & 1 & 0 \\ 0 & 0 & 0 \\ 0 & 0 & 0 \end{pmatrix}, \quad \mathcal{B}_2 = \begin{pmatrix} 0 & 0 & 0 \\ 0 & 0 & 1 \\ 0 & 0 & 0 \end{pmatrix}, \quad \mathcal{B}_3 = \begin{pmatrix} 0 & 0 & 1 \\ 0 & 0 & 0 \\ 0 & 0 & 0 \end{pmatrix}$$

with commutation relations

$$(2.8) \qquad [\mathcal{B}_1, \mathcal{B}_2] = \mathcal{B}_3, \quad [\mathcal{B}_1, \mathcal{B}_3] = \Theta, \quad [\mathcal{B}_2, \mathcal{B}_3] = \Theta.$$

Here,

$$(2.9) \qquad B(x_1, x_2, x_3) = \exp x_3 \mathcal{B}_3 \exp x_2 \mathcal{B}_2 \exp x_1 \mathcal{B}_1.$$

The Lie algebra $L(H_R)$ is a *contraction* of $L(G_A \times R)$, [13], [19]. Rather than review the formal definition of contraction we shall say explicitly what we mean for this single example. Let $\epsilon \neq 0$ be a real parameter and define the elements $\mathcal{C}_j(\epsilon)$ of $L(G_A \times R)$ by

$$(2.10) \qquad \mathcal{C}_1(\epsilon) = \epsilon\, \mathcal{A}_1, \quad \mathcal{C}_2(\epsilon) = \mathcal{A}_2 - \mathcal{A}_3, \quad \mathcal{C}_3(\epsilon) = \epsilon\, \mathcal{A}_3.$$

Then $\{\mathcal{C}_1, \mathcal{C}_2, \mathcal{C}_3\}$ is a new basis for $L(G_A \times R)$ with commutation relations

$$(2.11) \qquad [\mathcal{C}_1, \mathcal{C}_2] = \epsilon \mathcal{C}_2 + \mathcal{C}_3, \quad [\mathcal{C}_1, \mathcal{C}_3] = \Theta, \quad [\mathcal{C}_2, \mathcal{C}_3] = \Theta.$$

Thus $[\mathcal{C}_j(\epsilon), \mathcal{C}_k(\epsilon)] = \sum_{\ell=1}^{3} c_{jk}^{\ell}(\epsilon)\mathcal{C}_\ell(\epsilon)$ where the $c_{jk}^{\ell}(\epsilon)$ are the *structure constants* of the Lie algebra $L(G_A \times R)$ with respect to this basis. Now pass to the limit in (2.11) as $\epsilon \to 0$. We see from (2.10) that in the limit the $\{\mathcal{C}_j\}$ no longer form a basis. However the structure constants on the right side of (2.11) have a well defined limit and, since they must still satisfy the antisymmetry and Jacobi identity conditions, [10], in the limit, the coefficients $c_{jk}^{\ell} = \lim_{\epsilon \to 0} c_{jk}^{\ell}(\epsilon)$ determine a Lie algebra, not necessarily isomorphic to $L(G_A \times R)$. It follows immediately from (2.11) that this new Lie algebra has commutation relations

$$(2.12) \qquad [\mathcal{C}_1(0), \mathcal{C}_2(0)] = \mathcal{C}_3(0), \quad [\mathcal{C}_1(0), \mathcal{C}_3(0)] = \Theta, \quad [\mathcal{C}_2(0), \mathcal{C}_3(0)] = \Theta.$$

Comparing (2.12) and (2.8) we see that with the identification $\mathcal{C}_1(0) \equiv -\mathcal{B}_2$, $\mathcal{C}_2(0) \equiv \mathcal{B}_1$, $\mathcal{C}_3(0) \equiv \mathcal{B}_3$, $L(H_R)$ is a contraction of $L(G_A \times R)$.

Now let us consider this contraction at the group level. For $\epsilon \neq 0$ an alternate parametrization of $G_A \times R$ is provided by the group elements

$$A[a, \beta, c] = \exp(c - a\beta)\mathcal{C}_3(\epsilon) \exp a\mathcal{C}_2(\epsilon) \exp(-\beta)\mathcal{C}_1(\epsilon)$$

$$(2.13) \qquad = \exp \epsilon(c - a\beta)\mathcal{A}_3 \exp a(\mathcal{A}_2 - \mathcal{A}_3) \exp(-\epsilon\beta)\mathcal{A}_1$$

$$= \begin{pmatrix} e^{-\epsilon\beta} & a & 0 \\ 0 & 1 & 0 \\ 0 & 0 & e^{\epsilon(c-a\beta)-a} \end{pmatrix} = A\left(e^{-\epsilon\beta}, a, \epsilon(c - a\beta) - a\right).$$

In terms of the group parameters $[a, \beta, c]$ the group multiplication law for $G_A \times R$ is

$$[a, \beta, c][a', \beta', c'] = [a + a'e^{-\epsilon\beta}, \beta + \beta',$$
$$(2.14) \qquad c + c' + a\beta' + \frac{a'}{\epsilon}(e^{-\epsilon\beta} - 1) + a'e^{-\epsilon\beta}\beta + a'\beta'(e^{-\epsilon\beta} - 1)].$$

In the limit as $\epsilon \to 0$, (2.14) becomes

$$(2.15) \qquad [a, \beta, c][a', \beta', c'] = [a + a', \beta + \beta', c + c' + a\beta'],$$

which is the multiplication law for H_R with $x_1 = a$, $x_2 = \beta$, $x_3 = c$.

The action of $G_A \times R$ associated with the wideband ambiguity function is the unitary representation $\mathbf{L}^\lambda(a, b, c)$ acting on the Hilbert space $L_2(R)$ of Lebesgue square integrable functions f on the real line:

$$(2.16) \qquad \mathbf{L}^\lambda(a, b, c)f(t) = a^{-\frac{1}{2}}e^{2\pi i\lambda c}f\left(\frac{t + b}{a}\right).$$

Here λ is a real parameter, (a, b, c) are the coordinates (2.1) on $G_A \times R$ and the operators $\mathbf{L}^\lambda$ satisfy the group representation property

$$(2.17) \qquad \mathbf{L}^\lambda(a, b, c)\mathbf{L}^\lambda(a', b', c') = \mathbf{L}^\lambda(aa', ab' + b, c + c').$$

Further, we have

(2.18) $$\langle \mathbf{L}^\lambda(a,b,c)f_1, \mathbf{L}^\lambda(a,b,c)f_2\rangle = \langle f_1, f_2\rangle.$$

The Lie algebra of generalized Lie derivatives associated with the representation (2.16) has the basis

(2.19) $$A_1 = -t\frac{d}{dt} - \frac{1}{2}, \quad A_2 = \frac{d}{dt}, \quad A_3 = 2\pi i\lambda.$$

Indeed

(2.20) $$[A_1, A_2] = A_2, \quad [A_1, A_3] = [A_2, A_3] = O$$

where O is the zero operator, and

$$\mathbf{L}^\lambda(a,b,c) = \exp cA_3 \exp bA_2 \exp \ln a A_1.$$

As stated earlier, the action of H_R associated with the narrowband ambiguity function is the unitary irreducible representation $\mathbf{T}^1$ acting on the Hilbert space $L_2(R)$, defined by the unitary operators $\mathbf{T}^1[x_1, x_2, x_3]$:

(2.21) $$\mathbf{T}^1[x_1, x_2, x_3]g(t) = e^{2\pi i(x_3 + tx_2)}g(t + x_1), \quad g \in L_2(R).$$

These operators satisfy the group property

(2.22) $$\mathbf{T}^1[x_1, x_2, x_3]\mathbf{T}^1[y_1, y_2, y_3] = \mathbf{T}^1[x_1 + y_1, x_2 + y_2, x_3 + y_3 + x_1 y_2].$$

Moreover, the associated Lie algebra of generalized Lie derivatives has the basis

(2.23) $$B_1 = \frac{d}{dt}, \quad B_2 = 2\pi it, \quad B_3 = 2\pi i$$

with commutation relations

(2.24) $$[B_1, B_2] = B_3, \quad [B_1, B_3] = [B_2, B_3] = O.$$

Now we will show how to relate the representations $\mathbf{L}^\lambda$ and $\mathbf{T}^1$ through the group contraction procedure. The method is most easily understood at the Lie algebra level. In the Hilbert space associated with $\mathbf{L}^\lambda$ we express each $f \in L_2(R)$ in the form $f(y) = e^{2\pi i\lambda t}g(t)$. Clearly $g \in L_2(R)$ and we can transfer the action of the operators $\mathbf{L}^\lambda(a,b,c)$ to the $g(t)$ via the unitarily equivalent action $\mathbf{K}^\lambda(a,b,c) = e^{-2\pi i\lambda t}\mathbf{L}^\lambda(a,b,c)e^{2\pi i\lambda t}$:

(2.25) $$\mathbf{K}^\lambda(a,b,c)g(t) = a^{-\frac{1}{2}}e^{2\pi i\lambda c}e^{2\pi i\lambda[t(1-a)+b]/a}g\left(\frac{t+b}{a}\right).$$

The associated Lie algebra action is defined by the operators

$$(2.26) \qquad A_1' = -t\frac{d}{dt} - \frac{1}{2} - 2\pi i \lambda, \quad A_2' = \frac{d}{dt} + 2\pi i \lambda, \quad A_3' = 2\pi i \lambda,$$

which satisfy the commutation relations (2.20) again. Now, in analogy with (2.10), we construct a one-parameter family of bases $\{C_1, C_2, C_3\}$ for this Lie algebra:

$$(2.27) \qquad C_1(\epsilon) = \epsilon A_1' = -\epsilon t \frac{d}{dt} - \frac{\epsilon}{2} - 2\pi i \epsilon \lambda(\epsilon) t$$
$$C_2(\epsilon) = A_2' - A_3' = \frac{d}{dt}, \quad C_3(\epsilon) = \epsilon A_3' = 2\pi i \epsilon \lambda(\epsilon).$$

This family of bases satisfies the commutation relations (2.11). The only nonzero commutator is

$$[C_1(\epsilon), C_2(\epsilon)] = \epsilon C_2(\epsilon) + C_3(\epsilon).$$

Note that we have allowed the parameter λ to vary with ϵ. Indeed we choose $\lambda(\epsilon) = \epsilon^{-1}$. Then in the limit as $\epsilon \to 0$ we obtain the result

$$(2.28) \qquad C_1(0) = -2\pi i t = -B_2, \quad C_2(0) = \frac{d}{dt} = B_1, \quad C_3(0) = 2\pi i = B_3,$$

i.e., the Lie algebra version of the representation $\mathbf{T}^1$ of H_R.

At the group level, we see from (2.13) that for $\epsilon \neq 0$ we have the representation $\mathbf{K}^{\epsilon^{-1}}$ of $G_A \times R$:

$$(2.29) \qquad \mathbf{K}^{\epsilon^{-1}}[a, \beta, c] = \exp[c - a\beta]C_3(\epsilon) \exp aC_2(\epsilon) \exp -\beta C_1(\epsilon)$$
$$\mathbf{K}^{\epsilon^{-1}}[a, \beta, c]g(t) = e^{2\pi i(c - a\beta + \epsilon\beta/2)} e^{\frac{2\pi i}{\epsilon}(e^{\epsilon\beta} - 1)(t + a)} g\left((t + a)e^{\epsilon\beta}\right).$$

For $\epsilon \neq 0$, $\mathbf{K}^{\epsilon^{-1}}$ is unitary equivalent to the representation $\mathbf{L}^{\epsilon^{-1}}$ of $G_A \times R$. In the limit $\epsilon \to 0$ it becomes

$$\mathbf{K}[a, \beta, c]g(t) = e^{2\pi i c} e^{2\pi i \beta t} g(t + a),$$

in agreement with the operator $\mathbf{T}^1[x_1, x_2, x_3]$, (2.21), associated with H_R for $x_1 = a$, $x_2 = \beta$, $x_3 = c$.

It follows from these results that the narrowband cross-ambiguity function can be obtained from the wideband cross-ambiguity function by taking the limit

$$(2.30) \qquad \lim_{\epsilon \to 0} e^{-i(a\beta + a/\epsilon)} \langle \mathbf{L}(e^{-\epsilon\beta}, a)f_1^\epsilon, f_2^\epsilon \rangle = \langle \mathbf{T}^1(a, \beta)g_1, g_2 \rangle$$

where $\mathbf{L}$ and $\mathbf{T}^1$ are given by (1.7) and (1.11), respectively. Here, $f_j^\epsilon(t) = e^{it/\epsilon} g_j(t)$. Moreover, for $\epsilon > 0$ sufficiently small we have

$$|\langle \mathbf{L}(e^{-\epsilon\beta}, a)f_1^\epsilon, f_2^\epsilon \rangle| \approx |\langle \mathbf{T}^1(a, \beta)g_1, g_2 \rangle| = |\langle \mathbf{T}^1(a, \beta)f_1^\epsilon, f_2^\epsilon \rangle|.$$

3. Decomposition of the regular representation. Recall that if $f(a, b)$ is a function on the group G_A, square integrable with respect to the right invariant measure $d_r A = \frac{da\,db}{a}$, then f can be expanded in the form

$$(3.1) \qquad f(a, b) = \sum_{n=0}^{\infty} \langle \tilde{e}_n, \mathbf{L}(a, b) s_n \rangle$$

where $\{s_n\}$ is an orthonormal basis for $L_2(R)$. Further, if s_n is continuously differentiable,

$$
\begin{aligned}
\tilde{e}_n(t) &= \frac{-i}{2\pi} \int_{G_A} f(a', b') \mathbf{L}(a', b') \frac{ds_n}{dt}(t)\, d_\ell A \\
(3.2) \qquad &= \frac{-i}{2\pi} \int_{-\infty}^{\infty} \int_0^{\infty} \frac{f(a', b')}{\sqrt{a'}} \frac{ds_n}{dt}\left(\frac{t+b'}{a'}\right) \frac{da'\,db'}{(a')^2}
\end{aligned}
$$

where $d_\ell A = \frac{da'\,db'}{(a')^2}$ is the left invariant measure on G_A. Using the concept of group contraction, we will show how (3.1) goes to the expansion formula (1.14) for the group H_R, in an appropriate limit. Strictly speaking, we should consider instead the expansion formula for functions $f(a^\circ, b^\circ, c^\circ)$ on the group $G_A \times R$. Since the unitary irreducible representations of R are one-dimensional, it is sufficient to consider functions of the form $f_\lambda[a^\circ, b^\circ, c^\circ] = f_\lambda(a^\circ, b^\circ) e^{2\pi i \lambda c^\circ}$ for λ a real parameter. The right invariant measure on $G_A \times R$ is $da^\circ\,db^\circ\,dc^\circ/a^\circ$, and the left invariant measure is $da^\circ\,db^\circ\,dc^\circ/(a^\circ)^2$. Introducing the coordinates $[a', \beta', c']$, (2.13)

$$a^\circ = e^{-\epsilon\beta'}, \quad b^\circ = a', \quad c^\circ = \epsilon(c' - a\beta') - a',$$

we find that the right and left invariant measures are, respectively,

$$(3.3) \qquad \epsilon^2 d\beta'\,da'\,dc', \qquad \epsilon^2 e^{\epsilon\beta'} d\beta'\,da'\,dc'.$$

Performing the integration on c' from $-\infty$ to $+\infty$ we cancel a factor $\epsilon\,dc'$ in (3.3). Finally we set $s_n(t) \equiv s_n^{\lambda^{-1}}(t) = e^{2\pi i \lambda t} g_n(t)$ and $\lambda = \epsilon^{-1}$. Then $\tilde{e}_n(t)$ becomes

$$(3.4)$$

$$
\begin{aligned}
\tilde{e}_n^{(\epsilon)}(t) = \frac{-i}{2\pi} \int_{-\infty}^{\infty} \int_{-\infty}^{\infty} f_{\epsilon^{-1}}(e^{-\epsilon\beta'}, a') e^{\epsilon\beta'/2} e^{-2\pi i(a'\beta' + a'/\epsilon)}. \\
\cdot \left[\frac{2\pi i}{\epsilon} g_n\left(\frac{t+a'}{e^{-\epsilon\beta'}}\right) + \frac{dg_n}{dt}\left(\frac{t+a'}{e^{-\epsilon\beta'}}\right) \right] e^{\frac{2\pi i}{\epsilon}(t+a')e^{\epsilon\beta'}} \epsilon e^{2\epsilon\beta'}\, d\beta'\,da'
\end{aligned}
$$

and

$$(3.5) \qquad \langle \tilde{e}_n^\epsilon, \mathbf{L}(e^{-\epsilon\beta}, a) s_n^\epsilon \rangle = \int_{-\infty}^{\infty} \tilde{e}_n^\epsilon(t) e^{\epsilon\beta/2} \overline{g}_n\left(\frac{t+a}{e^{-\epsilon\beta}}\right) e^{-\frac{2\pi i}{\epsilon}(t+a)e^{\epsilon\beta}}\, dt$$

with

$$(3.6) \qquad f_{\epsilon^{-1}}(e^{-\epsilon\beta}, a) e^{-2\pi i(a\beta + a/\epsilon)} = \sum_{n=0}^{\infty} \langle \tilde{e}_n^\epsilon, \mathbf{L}(e^{-\epsilon\beta}, a) s_n^\epsilon \rangle.$$

Now we multiply both sides of (3.6) by the factor $e^{2\pi i(a\beta+a/\epsilon)}$ and pass to the limit as $\epsilon \to 0$. With the assumption that $\lim_{\epsilon \to 0} f_{\epsilon^{-1}}(e^{-\epsilon\beta}, a) = F(a, \beta)$ where F is Lebesgue square integrable with respect to the measure $d\beta\, da$, we find

$$F(a, \beta) = \sum_{n=0}^{\infty} \langle e_n, \mathbf{T}^1(a, \beta)g_n \rangle$$

where

$$e_n(t) = \int_{-\infty}^{\infty} \int_{-\infty}^{\infty} F(a', \beta')e^{2\pi i\beta' t}g_n(t + a')\, da'\, d\beta',$$

in agreement with the expansion formula (1.14).

We note that from the point of view of group representation theory, expansions (1.8), (1.10) and (1.14), can be expressed as traces of operators, i.e., in a basis independent fashion, [9]. We have followed Naparst's choice [10] of a signal basis for the wideband case in our derivation. However, this choice is not essential for the validity of the proofs.

REFERENCES

[1] L. AUSLANDER AND I. GERTNER, *Wide-band ambiguity functions and $a \cdot x + b$ group*, Signal Processing, Part I: Signal Processing Theory, Vol 22. IMA Volumes in Mathematics and its Applications, L. Auslander, T. Kailath and S. Mitter, eds., Springer-Verlag, New York (1990), pp. 1–12.

[2] L. AUSLANDER AND R. TOLIMIERI, *Radar ambiguity function and group theory*, SIAM J. Math. Anal., 16, No. 3 (1985), pp. 577–601.

[3] R. E. BLAHUT, *Theory of remote surveillance algorithms*, in RADAR and SONAR, Part 1, IMA Volumes in Mathematics and its Applications, Springer-Verlag, New York (1991).

[4] C. E. COOK AND M. BERNFELD, *Radar Signals*, Academic Press, New York (1967).

[5] S. GAAL, *Linear Analysis and Representation Theory*, Springer-Verlag, New York, 1973.

[6] I. KHALIL, *Sur l'analyse harmonique du groupe affine de la droite*, Studia Mathematica, LI(2) (1974), pp. 139–167.

[7] A. KOHARI, *Harmonic analysis on the group of linear transformations of the straight line*, Proceedings of the Japanese Academy, 37 (1961), pp. 250–254.

[8] W. MILLER, JR., *Symmetry Groups and their Applications*, Academic Press, New York, 1972.

[9] W. MILLER, JR., *Topics in harmonic analysis with applications to Radar and Sonar*, in RADAR and SONAR, Part 1, IMA Volumes in Mathematics and its Applications, Springer-Verlag, New York (1991).

[10] H. NAPARST, *Radar signal choice and processing for a dense target environment*, Signal Processing, Part II: Control Theory and Applications, Vol 23, IMA Volumes in Mathematics and its Applications, F.A. Grünbaum, J.W. Helton and P. Khargonekar, eds., Springer-Verlag, New York (1990), pp. 293–319.

[11] A. NAYLOR AND G. SELL, *Linear Operator Theory in Engineering and Science*, Holt, New York, 1971.

[12] F. RIESZ AND B. SZ.-NAGY, *Functional Analysis*, English Transl. Ungar, New York, 1955.

[13] E. J. SALETAN, *Contractions of Lie groups*, J. Math. Phys., 2 (1961), pp. 1–21.

[14] W. SCHEMPP, *Radar ambiguity functions, the Heisenberg group, and holomorphic theta series*, Proc. Amer. Math. Soc., 92 (1984), pp. 103–110.

[15] W. SCHEMPP, *Harmonic analysis on the Heisenberg nilpotent Lie group with applications to signal theory*, Longman Scientific and technical, Pitman Research Notes in Mathematical Sciences, 147, Harlow, Essex, UK (1986).

[16] D. A. SWICK, *An ambiguity function independent of assumption about bandwidth and carrier frequency*, NRL Report 6471, Naval Research Laboratory, Washington, D.C., December (1966).

[17] D. A. SWICK, *A Review of Wideband Ambiguity Functions*, NRL Report 6994, Naval Research Laboratory, Washington, D.C., December (1969).

[18] N. JA. VILENKIN, *Special Functions and the Theory of Group Representations*, American Mathematical Society, Providence, Rhode Island (1966).

[19] E. P. WIGNER AND E. INÖNU, *On the contractions of groups and their representations*, Proc. Nat. Acad. Sci. U. S., 39 (1953), pp. 510–524.

[20] C. H. WILCOX, *The Synthesis Problem for Radar Ambiguity Functions*, MRC Technical Summary Report 157, Mathematics Research Center, United States Army, University of Wisconsin, Madison, Wisconsin, April (1960), *Also in RADAR and SONAR, Part 1, by R. E. Blahut, W. Miller and C. H. Wilcox, IMA Volumes in Mathematics and its Applications, Springer-Verlag, New York, 1991,*.

WIDEBAND APPROXIMATION AND WAVELET TRANSFORM

PETER MAASS*

Abstract. These notes illustrate how results from wavelet theory translate into radar language, treating the cross-correlation functions and the radar inverstion problem. Representation theory on the affine group and wavelets are composed as two methods to obtain an explicit inversion formula for the wideband approximation.

Key words. wavelet transfrom, wideband radar, cross correlation

AMS(MOS)subject classification. 42A38

1. Introduction. It is well known that "somehow" there exists a relation between the wideband approximation of radar echoes and the wavelet transform. These notes are aimed to clarify "somehow" and to illustrate how results from wavelet theory translate into the radar language. The investigation centers around the cross-correlation function and the consequences on the radar inversion problem.

We do not consider discretizations: hence the name "wavelet transform" here refers to the continuous wavelet transform. This is usually introduced by group theory on the affine group. Some basic relations to the radar problem can be found (see the lecture notes of W. Miller [2]). For an excellent paper on this subject from the radar engineers point of view see [5]. On the other hand the wavelet transform can be defined as an integral operator. The investigation of its analytical properties overcomes one restriction of the algebraic approach, namely the "admissibility condition" for the signals. The consequences for radar inversion procedures are discussed in Section 4.

These notes serve a second purpose. Representation theory on the affine group was the major tool used by Naparst [3] to obtain an explicit inversion formula for the wideband approximation. Both approaches, by wavelets or purely group-theoretic, therefore have the same basis and yield similar results. For a comparison see Section 3.

2. The basic relations. First of all we have to define the wideband approximation of a radar echo in a dense target environment. The target density is described by a square integrable function $D(x, y)$ in the delay-Doppler plane. An emitted signal $\psi(t)$ is reflected by this density and its echo $e(t)$ is recorded. The echo is time delayed and transformed according to the Doppler effect. Then the wideband approximation is given by

$$e(t) = \int \int D(x, y) \sqrt{y} \, \psi(\, y(t - x) \,) \, dx dy \ .$$

We want to express the echo in terms of a wavelet transform. The wavelet transform is introduced as an integral operator whose kernel depends on the so called "analyzing

---

* Fachbereich Mathematik, Technische Universität Berlin, 1ooo Berlin 12, West Germany.

wavelet ψ", $I\!\!R_+^2 = \{(x,y)|x,y \in I\!\!R ,y > 0\}$:

$$L_\psi \;:\; L_2(I\!\!R) \;\longrightarrow\; L_2(I\!\!R_+^2, \frac{dxdy}{y^2})$$

$$f \;\longrightarrow\; \frac{1}{\sqrt{y}} \int f(t)\, \psi(\frac{t-x}{y})dt \quad .$$

One of the basic advantages of the wavelet transform is its capability of encoding local time and frequency information simultaneously. A very readable survey on continuous and discrete wavelet transforms is contained in [4]. The wideband approximation is most conveniently expressed with the help of the adjoint wavelet transform:

$$L_\psi^* \;:\; L_2(I\!\!R_+^2, \tfrac{dxdy}{y^2}) \;\longrightarrow\; L_2(I\!\!R)$$

$$\tilde{D} \;\longrightarrow\; \int\int \tilde{D}(x,y)\, \frac{1}{\sqrt{y}}\, \psi(\frac{t-x}{y})\, \frac{dxdy}{y^2} \quad .$$

A simple change of coordinates $y \to 1/y$ yields the basic relation between radar echoes and the wavelet transform :

$$\boxed{L_\psi^* \tilde{D} \;=\; e \;,\;\; \tilde{D}(x,y) = D(x,1/y)}$$

Usually these transforms are introduced via group theory because the wavelet transform is governed by the action of the affine group on L_2 functions on the real line. For our purposes we could work without any group theory but we also want to compare the wavelet approach with the results of [3]. There a purely group-theoretic approach, based on representation theory on the affine group, leads to inversion formulae for the wideband approximation. Therefore both methods are closely related and their similarities and differences will be discussed in the next chapter after the introduction of the necessary notation.

G_A denotes the affine group, "$\frac{t-x}{y}$", of translations by x and dilations by y. The left Haar measure equals $d\mu(x,y) = dxdy/y^2$, which explains the peculiar weight function in the definition of the wavelet transform as an integral operator. The affine group has a unitary representation U on the space of L_2 functions on the real line:

$$U(x,y)f(\cdot) = \frac{1}{\sqrt{y}}\, f(\frac{\cdot - x}{y}) \quad .$$

Given an admissible function ψ, (an explanation of admissibility is given in the next chapter) the wavelet transform and the adjoint wavelet transform translate into simple scalar products :

$$L_\psi f(x,y) = \langle f, U(x,y)\psi \rangle_{L_2(I\!\!R)} \quad ,$$
$$L_\psi^* D(t) = \langle D, U(x,y)\psi \rangle_{L_2(G_A)} \quad .$$

In both cases the function to be transformed is integrated against $U(x,y)\psi(t)$: in the first equation one has to integrate over $I\!\!R$, in the second over G_A. In other words the wavelet transform is an isometry between the appropriate L_2 spaces.

Given a function $D \in L_2(G_A)$ its Fourier transform with respect to the representation U acts on a function $f \in L_2(I\!\!R)$ according to

$$(\mathcal{F}D)f \;=\; \int_{G_A} D(x,y)\, U(x,y)f\, d_\mu(x,y) \quad .$$

With these definitions the wideband approximation of a radar echo reduces to the following simple equation :

$$(\mathcal{F}\tilde{D})\psi \;=\; e$$

3. Vocabulary. Starting from the basic equivalent formulations for the wideband approximation of a radar echo we can express almost everything connected with wideband radar in the language of wavelets or of group theory. A preliminary vocabulary follows.

Radar	Affine Group G_A	Wavelet
wideband echo	Fourier transform	$(\mathbf{L}_\psi^* \mathbf{D})(\mathbf{t})$
ambiguity	matrix element of U	$(\mathbf{L}_{\psi_n}\psi_m)(\mathbf{x},\mathbf{y})$
cross-correlation	$\langle \mathcal{F}(D)\psi_n, U(x,y)\psi_m \rangle$	$(\mathbf{L}_\psi\mathbf{L}_\psi^*\mathbf{D})(\mathbf{x},\mathbf{y})$
Inversion formulae	Peter-Weyl Theorem	$\mathbf{Id} = \mathbf{L}_\psi^*\mathbf{L}_\psi$
	trace form of Peter-Weyl Theorem	$\mathbf{L}_\psi\mathbf{L}_\psi^*$ projection onto $range(L_\psi)$
restrictions for ψ	$\hat{\psi}(\omega) = \omega\hat{\phi}(\omega)$	ψ admissible

Any entry in this table exhibits interesting relations to radar applications and needs further investigation. In the following we shall concentrate on a discussion of the inversion formulae but we cannot resist remarking on some of the other entries. As we have seen in the last chapter the echo of a signal ψ reflected by a target density D corresponds to applying the adjoint wavelet transform L_ψ^* with basic wavelet ψ to D. Moreover cross-correlating the echo with ψ amounts to computing the wavelet transform L_ψ of e. The well-known projection property of the wavelet transform translates into the statement that the outcome of the cross-correlation procedure is the projection of the searched-for density D onto the range of L_ψ. A characterization of this range shows which part of D can be recognized by cross correlation. A similar argument applies for the ambiguity function which can be expressed in different ways in terms of wavelet transforms. The expression given here characterizes the ambiguity function as a very special element of $range(L_\psi)$.

The inversion formulae stated in the column "Affine Group" refer to the work of Naparst [3]. The elegant trace form of the Peter-Weyl theorem, sometimes called formula of Khalil, allows an explicit reconstruction of D by sending out a family of signals $\{\psi_n\}$ and cross correlating the echoes e_n with a set of functions $\{\phi_n\}$. More precisely the algorithm proceeds as follows:

- choose a basis $\{\phi_n\}$ of $L_2(I\!R)$
- construct signals by $\hat{\psi}(\omega) = \omega\hat{\phi}(\omega)$
- record the echoes $\{e_n\}$ of $\{\psi_n\}$
- cross correlate e_n with ϕ_n
- summing the cross-correlations reproduces D

The class of signals allowed in this process are forced therefore to obey the following condition

$$\int \frac{|\hat{\psi}_n(\omega)|^2}{\omega^2} d\omega < \infty \quad .$$

For a moment let us consider what can be done with a single signal ψ. Then the wavelet inversion formulae, i.e. simple cross correlation, reconstructs the projection of D onto $range(L_\psi)$, provided the signal fulfills the following (normalized) **admissability condition** :

$$\int \frac{\hat{\psi}(\omega)^2}{|\omega|} d\omega = 1 \quad .$$

Clearly the signals used by Naparst are also admissible wavelets. Both conditions force the signals to fulfill

$$\hat{\psi}(0) = \int \psi(t)dt = 0 \quad ,$$

Only signals with zero mean value can be used. This admissibility condition stems from the group-theoretic derivation of the wavelet transform and the central issue of the next section is how to bypass this restriction. Straightforward analytic manipulations are necessary in order to allow the use of arbitrary signals ϕ without loosing the ability to compute the cross-correlation $L_\psi L_\psi^* D$ with an admissible ψ.

4. Analytic properties of the wavelet transform. In the last section different possibilities to reconstruct the density D were discussed. Both procedures suffer from severe restrictions concerning the admissibility of signals: only signals with zero mean are suitable. This restriction was necessary in order to analyze the cross-correlation function in terms of wavelets or representation theory on the affine group.

Now we wish to allow an arbitrary signal ϕ. Starting from the echo e produced by ϕ we want to calculate the cross-correlation function as if a signal ψ fulfilling the admissibility condition were used. We depend on the results proved by Rieder [4]. The following statement is the simplest form of Theorems 3.3 and 3.4 of [4].

Corollary. *Suppose* ϕ , $\phi' \in L_2(I\!R)$, ϕ' *denotes the (generalized) derivative of* ϕ. *Then* $\psi = \phi'$ *is admissible if*

$$\hat{\phi}(0) = c \neq 0 \, , \, \psi := \phi' \Rightarrow \lim_{y \to 0} y^{-3/2}(L_\psi f)(\cdot, y) = cf'(\cdot) \quad .$$

The second statement says that the wavelet transform $L_\psi f$, $y > 0$, is a regularization of the the first derivative of f under certain conditions. This is the reason for the power of the wavelet transform in edge detection and pattern recognition. We add some simple rules describing how taking derivatives intertwines with L_ψ and L_ψ^*. In an abuse of notation we write L_ϕ for the integral operator defined in Section 1 even if ϕ is not admissible.

Lemma. *Suppose* $\phi, \phi' \in L_2(\mathbb{R})$, f, D *denote continuous functions with compact support in* $\mathbb{R}$, *resp.* $\mathbb{R}_+^2$, $\psi := \phi'$. *Then*

$$\frac{d}{dx}(L_\phi f)(x,y) = (\frac{-1}{y})(L_\psi f)(x,y) \ , \quad (L_\psi f)(x,y) = -y(L_\phi f')(x,y) \ ,$$

$$\frac{d}{dt}(L_\phi^* D)(t) = L_\psi^*(\frac{1}{y}D)(t) \ ,$$

$$\frac{d}{dx}(L_\psi L_\phi^* \tilde{D})(x,y) = (L_\psi L_\psi^*)(\frac{1}{y}\tilde{D})(x,y) \ .$$

The proof of the first part follows from the definition of the wavelet transform as an integral operator and by justifying the interchange of integration and differentiation. The second part is proved by applying each formula of the first part once. □

The algorithm now requires a differentiable signal, but no requirement for zero mean is needed, :
  - use a differentiable signal ϕ
  - cross-correlate the echo $e(t) = (L_\phi^* \tilde{D})(t)$ with $\psi = \phi'$
  - calculate $\frac{\partial}{\partial x}$ in a regularized way, e.g. apply a suitable wavelet transform. This procedure results in an approximation of the projection of $(1/y)\tilde{D}$ onto $range(L_\psi)$.

We can play around with the formulae of the lemma and derive different expressions for the wanted quantity $(L_\psi L_\psi^*)(\frac{1}{y}\tilde{D})(x,y)$, the expression given in lemma seems to be the one with the greatest potential for applications :

$$(L_\psi L_\psi^*)(\frac{1}{y}\tilde{D})(x,y) \ = \ (-y)\frac{d^2}{dx^2}(L_\phi e)(x,y) \ = \ \frac{1}{y}(L_{\psi'}e)(x,y) \ .$$

If we want to reconstruct D itself we have to go back to group theory. Suppose we send out a family of signals $\{\phi_n\}$, such that the corresponding $\{\psi_n\}$ form a basis for $L_2(\mathbb{R})$, then by the above procedure we can calculate

$$(L_{\psi_m} L_{\psi_n}^*)(\frac{1}{y}\tilde{D})(0,1) \ = \ \langle D, L_{\psi_n}\psi_m\rangle_{G_A} \ .$$

By our vocabulary this is a matrix element of the representation of the affine group and the Peter-Weyl theorem tells us how to reconstruct the density $D(x,y)$ itself.

5. Conclusions. As we have seen in the previous sections the wavelet transform is suitable for expressing almost any quantity connected with wideband radar. Theoretic results on the wavelet transform translate into interpretable and useful results for radar applications. We were mainly concerned with the cross-correlation function, which is the projection of D onto the range of L_ψ. Regarding the radar inversion process it seems to be advantageous not to cross correlate with the signal ϕ itself, correlating instead with a related signal ψ allows to use a broader class of signals and to obtain a total reconstruction of D by a series expansion. However the application hinges on the ability to generate arbitrary signals.

REFERENCES

[1] C.E. HEIL AND D.F. WALNUT, *Continuous and discrete wavelet transforms*, SIAM Review
    Vol.31, No. 4 (1989), pp. 628-666
[2] W. MILLER, *Topics in harmonic analysis with applications to radar and sonar*, in Radar and
    Sonar, A. Grünbaum, M. Bernfeld, and R. Blahut, eds., IMA Volume in Mathematics
    and its Applications, Springer-Verlag, New York, (1991).
[3] H. NAPARST, *Radar signal choice and processing for a dense target enviroment*, Ph.D. thesis,
    University of California, Berkeley, (1988)
[4] A. RIEDER, *Approximationseigenschaften der Wavelet Transformation*, Ph.D. thesis, Technische
    Universität Berlin, (1990)
[5] P. BERTRAND AND J. BERTRAND, *Time-frequency representations of broad band signals*,
    La Recherche Aerospatiale 1985-5, (1985), pp. 277-283.

PROBLEMS IN STATIONARITY
FOR LARGE ACOUSTIC ARRAYS

GERALD L. MOHNKERN*

Abstract. As acoustic receiving arrays become larger, several problems in the minimum variance distortionless response (MVDR) algorithm for frequency-domain adaptive beamforming become apparent. As the physical dimensions of arrays become larger, the length of the discrete Fourier transform must become larger to assure coherence for the increased delays encountered between sensors. At the same time, the number of discrete Fourier transforms required to obtain a full rank matrix matrix increases with the number of sensors. For a filled line array, this seems to imply that the amount of data which must be accumulated to obtain an acceptable estimate of the noise field for MVDR increases as the square of the length of the array. The problem is exacerbater because in realistic noise fields adding more sensors increases the range of eigenvalues, decreasing the numerical stability of the cross-spectral density matrix. Furthermore, a longer array has more spatial resolution, so that smaller motions of interferences and targets become significant nonstationarities. All of these make stationarity of the noise field a major problem in designing the processing for large arrays. Several potential solutions will be discussed.

1. Introduction. There has recently been a renewed interest in large acoustic receiving arrays for undersea applications. This interest has been driven by predictions of quieter submarine targets, because of improved tooling technology and also because of the wider availability of quiet diesel submarines to many countries.

Frosh's law states that for propagation conditions which support convergence zone propagation, fewer sensors are required per unit area of coverage if they are deployed and processed as an array than if they are used as distributed sensors. Thus, for the ocean basins beyond the immediate continental shelves, an array, if it can be processed coherently, is more efficient than a distribution of independently processed sensors. Furthermore, an examination of shipping densities suggests that for large arrays operating in the frequency band where shipping dominates the noise field, there will be interstices between the shipping noise where the noise level will be significantly lower than the average noise. Exploitation of these interstices requires large array apertures.

Acoustic array sizes have been limited by the assumption that signals arrive at the array as plane waves. This assumption was necessitated by the lack of acoustic propagation models accurate enough to predict the received acoustic field over larger apertures and the lack of computation resources to process the larger number of field points if an appropriate model had been available. However, recent improvements in acoustic propagation modeling and in computer performance have made it feasible to process arrays with very large apertures.

Sensors for large arrays are expensive, even if fewer are required than for a distributed collection of sensors. Increasing the performance of a 10 sensor array by 3 dB requires adding only 10 more sensors, but increasing the performance of a 1000 sensor array by 3 dB requires adding 1000 sensors, so a 3dB improvement in a 1000 sensor array costs about 100 times as much as a 3 dB improvement in

---

*Defense Advanced Research Projects Agency, Arlington, VA 22209-2308.

a 10 sensor array. The unit cost of computation is decreasing while the unit cost of sensors is increasing, so it is prudent to achieve as much improvement in array performance as is possible by better processing instead of more sensors.

Adaptive noise suppression offers additional benefits for large arrays. For large arrays it will be increasingly difficult to maintain sensor positions on a precise grid, so peak sidelobe levels will increase. Furthermore, matched field processing (MFP), which applies acoustic models to array signals, tends to produce high sidelobes. Adaptive noise suppression can suppress interferences which fall on these sidelobes so that "beams" steered in the interstices of the shipping noise field will not be dominated by sidelobe noise. However, all types of adaptive beamforming require an estimate of the noise field, and estimating the noise field requires data. Therefore, stationarity of the noise field and the amount of data required to estimate the noise field limit performance.

2. Time Limitations. The most frequently discussed adaptive process is frequency domain minimum variance distortionless response (MVDR), so we consider it first. The discrete fourier transform (DFT) used to transform the data to the frequency domain must be at least twice as long as the maximum transit time across the array to assure that DFTs from all sensors include samples from the same point on a wavefront of a signal propagating from an arbitrary direction; ten times as long is the general rule of thumb used to control edge effects. the minimum number of sequential DFT samples required to achieve a nonsingular estimate of the sample correlation matrix (SCM) is equal to the number of sensors; twice as many samples as sensors is frequently used as a design criterion. If the array is a uniformly spaced line array with a half wavelength between sensors, the amount of data time required for a meaningful estimate of the noise field is therefore proportional to the square of the array length.

Furthermore, stationarity needs to be defined in meaningful terms. In frequency bands where shipping is the dominant source or noise, the noise field will appear nonstationary if one or more of the interfering ships moves by a significant fraction of a beamwidth during estimation of the SCM. Therefore, the time during which the noise field can be considered stationary decreases as array size increases.

As a specific example, consider a line array and environment with the following characteristics:

DFT length: 10 times propagation time
Sensor spacing: 1/2 wavelength
Number of DFT samples accumulated: 2 times number of sensors.

The closed-box line in Figure 1 shows the data time required to estimate the noise field for a frequency of 100 Hz as a function of the number of sensors in the array. The open boxes indicate the time required for an interference at a range of 50 Km to move across the broadside beam at 25 Km/Hr. The lines cross at about 50 sensors, indicating that the performance of narrowband MVDR will start to degrade if the example array contains more than 50 sensors. Other factors, such as sensor motion or internal waves may place even tighter limits on integration time.

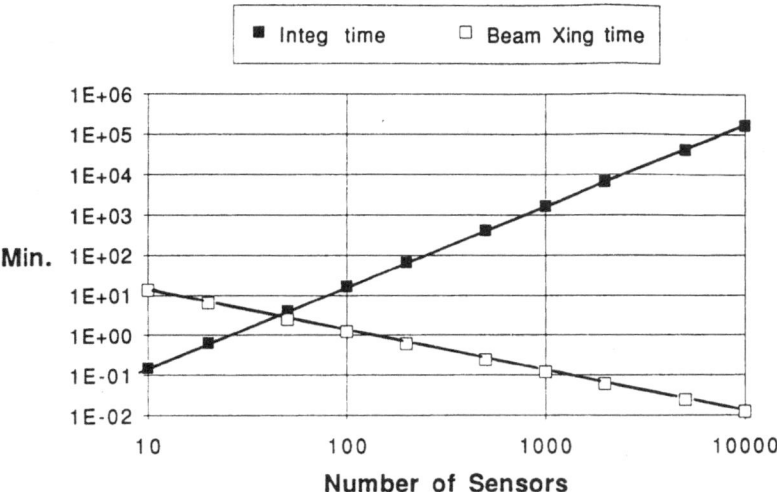

Figure 1: Integration Time and Beam Crossing Time for a Line
Array

3. Parallelism. The cost of almost all computation is decreasing, but the price/performance leaders are the parallel processors. An iPSC/860 with 128 processing units offers about 10 times the computation performance per dollar of a Cray YMP with 8 processing units. However, it is more difficult to fully utilize the processing power of the iPSC/860 because of its higher parallelism.

One approach to optimizing parallelism is to seek algorithms with a high degree of intrinsic parallelism, i.e. they can be split into a large number of parallel processes that do not require interprocess communication except to start and end the processes. The narrowband MVDR has a great deal of intrinsic parallelism, as illustrated in Figure 2. It is desirable that any algorithm replacing MVDR have a similar degree of intrinsic parallelism; lower forms of parallelism may offer similar performance, but usually at the expense of flexibility.

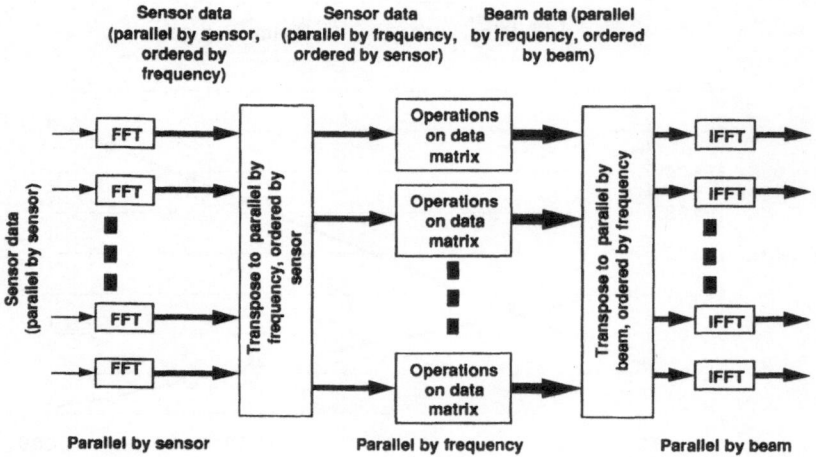

Figure 2: Intrinsic Parallelism of Narrowband MVDR Processing

4. Alternative Algorithms. If narrowband MVDR is replaced by a similar broadband process, there is no longer a minimum length of DFT to contend with, but the dimensions of the SCM must increase by a factor which is twice the number of samples in the acoustic propagation time across the array so that it can describe delays between all combinations of sensors. Therefore it requires the same order-of-magnitude estimation time. The computation costs are roughly proportional to the square of the dimension of the SCM, so the broadband process requires much more computation for little inprovement in sensitivity to nonstationarity. Computation is relatively cheap but not free.

Deterministic interference suppression by measuring the direction of the loudest interferences, estimating from a priori information the effect of that interference on each beam, and subtracting it out of each beam permits rapid suppression of strong interference, but is so sensitive to errors in the sensor locations and the propagation conditions that it is not a serious contendor.

Several approaches based on preprocessing part or all of the sensors have been proposed. The most widely discussed approach is beam-space adaptive beamforming, where a complete set of conventional beams are formed, and an SCM is computed for a limited subset of the beams. Beam space adaptive beamforming is highly efficient when there are more sensors than beams. However, if MFP is required there may be as many as 50,000 times as many "beams" as sensors, so it will be more advantageous to estimate the noise field in the sensor domain than in the beam domain.

A second preprocessing approach is to do the initial beamforming using subarrays. Each subarray is processed separately, probably using MVDR. Then, for each steering direction, each subarray is treated as a directional sensor and the subarray beams steered in the same "direction" are processed using either conventional or MVDR processing. The spacing of the directional sensors (subarrays) is sparse, so moiré patterns will be a problem if subarrays are equispaced.

Another preprocessing approach is to delay signals from all sensors for a "coarse" steering direction. Signals that propagate from the region of the coarse steering direction can then be processed using MVDR with a much shorter DFT. Signals arriving from other directions will not correlate between pairs of sensors in the SCM if their propagation time between the sensors is greater than the length of the DFT, so suppression of them will be limited to clusters of closely spaced sensors. However, signals near the coarse steering axis will benefit from full array gain, and interference near the coarse steering axis will be suppressed. It is sometimes useful to think of this approach as subarray processing, with the subarrays automatically selected based on their separation in the coarse steering direction.

In general, the interference to be suppressed are the strongest signals in the acoustic field, and are collectively approximated by the largest eigenvalues of the SCM, although the correspondence generally will not be one to one. This suggests that instead of using the inverse of the full rank SCM, one might be able to approximate the inverse from a truncated eigensystem. The MVDR filters would be computed from the approximate inverse, and applied to the original data so that small signals are not suppressed. The dynamic range of the eigensystem can be reduced by computing the singular value decomposition of the data matrix instead of the eigenvalues of the SCM. A difficult problem in analyzing the eigensystem approach is making meaningful estimates of the errors in the interference vectors of the truncated eigensystem.

5. Summary. Nonstationarity limits the accuracy of the representation of the noise field for adaptive processing of large arrays. There are several feasible approaches to alleviating the problem, but none of them stands out as being *the* solution. Each of the proposed solutions needs analytical and empirical evaluation, and there may be approaches not discussed in this paper which would prove more satisfactory. Evaluation criteria should be the suppression of the strongest interferences in a realistic, dynamic noise field, while not suppressing weak signals.

A SIEVE-CONSTRAINED MAXIMUM-LIKELIHOOD METHOD FOR TARGET IMAGING*

PIERRE MOULIN**, JOSEPH A. O'SULLIVAN† AND DONALD L. SNYDER†

Abstract. As radar systems become more versatile, there are fundamental limits to the achievable performance. When reconstructing target distributions from radar data, two limiting factors arising from statistical considerations are additive noise in the receiver and target reflections which are random. Another limiting factor imposed by data collection methods is due to a desire to reconstruct a continuous distribution of scatterers from a finite set of sampled data. In this paper, a diffuse target model is adopted, so reflections are assumed to be uncorrelated, complex-valued, and Gaussian distributed. The scattering function (which equals the expected intensity of the reflections) is the desired target image. The noise is assumed to be complex-valued white Gaussian noise. These statistical assumptions lead to a statistical model for the received data parameterized by the scattering function. A maximum likelihood estimation approach is adopted for estimating the scattering function. In order to address the estimation of a whole function, a method of sieves is adopted. The sieve set is composed of polynomial B-splines. Convergence of the estimates to the actual function as the received data set grows is shown using a measure of discrepancy based on the Kullback-Leibler divergence. An iterative algorithm for obtaining the maximum likelihood estimate in the sieve set is proposed.

1. Introduction.

As models for radar signal propagation and radar system design become more advanced, the opportunity for improving the performance of the systems using these models increases. In this paper, two issues affecting performance are addressed. The first is the statistical nature of the received signal. The statistical description adopted is valid for various radar systems under appropriate operating conditions [31,33]. This model has been used previously to derive algorithms for forming radar images using maximum likelihood (ML) methods [24,30]. The second issue addressed is the need for regularizing the estimates. The data are assumed to be samples of the received signal while the desired image is a whole function. The estimation problem may then be shown to be ill-posed [21,32] in the sense that small changes in the received data may lead to large changes in the estimated image. A method for regularizing the estimates based on the method of sieves [13] is proposed. The idea is to seek an estimate over a restricted subset of the parameter set. The size of this subset is controlled by a parameter μ called the mesh size. Assume that the estimate over the restricted subset exists. Then the mesh size should be large enough that the estimates are stabilized, yet small enough so that the estimated function is close to the true function. Our sieve sets consist of functions which are polynomial B-splines [27]. This choice allows us to incorporate nonnegativity constraints on the estimated function directly by constraining the coefficients in the

---

*The work described in this paper was supported by contract number N00014-86-K-0370 from the Office of Naval Research and by grant number MIP-8722463 from the National Science Foundation.

**Bell Communications Research, MRE 2M-393, 445 South Street, Morristown, NJ 07960.

†Electronic Systems and Signals Research Laboratory, Department of Electrical Engineering, Washington University, Saint Louis, MO 63130.

expansion to be nonnegative. A second advantage of this choice is the ability to characterize the resolution of the image estimate in terms of the size of the support of the splines. This motivates an interpretation of the algorithm as a multiresolution algorithm [21].

In order to quantify the image estimate quality, a discrepancy measure based on the Kullback-Leibler divergence is used. This measure is used to show convergence of the estimates to the true function as the size of the data set grows and the sieve is allowed to grow at an appropriate rate relative to the received data set. An optimal rate of growth of the sieve set may be obtained [21].

Our approach may be explained in a more general setting than that required for the radar problem discussed here. Let a complex random process be denoted by $\{z(u), u \in \mathbf{U}\}$, where $\mathbf{U}$ is a set in the d-dimensional Euclidean space $\mathbf{R}^d$. The second-order statistics take the form of an unknown, nonnegative real function $\{S(u), u \in \mathbf{U}\}$, and they are to be inferred from a set of complex-valued observations $\{r(t), t \in \mathbf{T}\}$. The observations are described by a linear transformation of $\{z(u), u \in \mathbf{U}\}$, corrupted by an additive noise $\{w(t), t \in \mathbf{T}\}$ with known statistics,

$$(1a) \qquad r(t) = \int_{\mathbf{U}} G(t, u) z(u) du + w(t), \qquad t \in \mathbf{T},$$

$$(1b) \qquad E[z(u)z^*(u')] = S(u)\delta(u - u'), \qquad u \in \mathbf{U}.$$

For the radar problem described in the following section, u is the pair of Doppler and delay coordinates (f, τ), and $\mathbf{U} \subset \mathbf{R}^2$. The kernel in (1a) is given by $G(t, u) = s_T(t - \tau)e^{j2\pi f(t-\tau/2)}$, where $s_T(t)$ is the transmitted radar signal [30]. For other problems this model is also valid. For a spectrum estimation problem, u is viewed as a frequency coordinate f, and $\mathbf{U} \subset \mathbf{R}$. The kernel in (1a) is then $G(t, u) = e^{j2\pi ft}$. The model in (1) may also be valid for wideband radar imaging. In that case, u is the pair of delay and time-scale coordinates (x, y), and $\mathbf{U} \subset \mathbf{R}^2$. The kernel is given by $G(t, u) = \sqrt{y}s_T(y(t - x))$ [23]. The equations in the following sections can be modified to this setting [22].

Conventional radar imaging or spectrum estimation techniques do not account for the statistical properties of the underlying process $\{z(u), u \in \mathbf{U}\}$. The estimation method is obtained from a purely deterministic analysis of the equation for the observations (1a). A typical approach is the following. First, some "reasonable" estimate for $\{z(u), u \in \mathbf{U}\}$ is produced by linear processing of the data. Next, the unknown intensity is estimated by taking the squared magnitude of this estimate. Under the statistical model considered here, these estimates have poor statistical properties, such as high variance. In radar imaging, various heuristic techniques have been proposed to improve the estimates [34 §10].

Our approach is to view the radar imaging problem as a statistical inference problem, and to use methods of statistical estimation theory to solve it. Although taking advantage of the information available in the form of a statistical model appears to be a natural strategy, this approach does not appear frequently in the radar literature. Statistical methods in spectrum estimation are much more common, but they generally assume more information about the process than is assumed here.

For instance, the process can consist of a superposition of sinusoids with unknown amplitude and phase. The classes of moving average (MA), autoregressive (AR), and ARMA models have also been all extensively studied [17]. When values of a set of correlation samples of the process are known, Burg's maximum-entropy method comes into consideration [6].

The approach taken here is more general in the sense that we do not wish to restrict our study to a particular type of process or a particular kind of prior information. In this sense, we simply view the whole function $\{S(u), u \in \mathbf{U}\}$ as an unknown parameter. We adopt the ML method to estimate it. So far, this approach has received only marginal attention in the radar and spectrum estimation literatures. Capon's method for spectrum estimation is sometimes called an ML method, but this has been recognized to be a misnomer [17]. Other existing ML methods assume a special form for the spectral density, such as ARMA or superposition of sinusoidal signals [5]. To our knowledge, the first use of ML for spectrum estimation in the general sense studied here is due to Chow and Grenander [8]. In radar imaging, research using this approach has been reported by Snyder, O'Sullivan, and Miller [30] and O'Sullivan and Snyder [24]. The method and estimation algorithm developed by the latter authors apply to spectrum estimation as well, and in that sense expands on the results in [20]. The application of these methods to the wideband radar imaging problem has not been investigated yet, although the results in [22] include it as a special case if the model in (1) is valid.

First, the radar imaging problem examined is described, including the discrete model used in the computations. A brief discussion of conventional imaging techniques motivates the need for sieves which are introduced in Section 4. The sieves are based on polynomial splines. An algorithm for obtaining the ML estimates in the sieves is proposed. In Section 5, a measure of discrepancy for estimates of the scattering function is introduced, which is based on the Kullback-Leibler divergence of the resulting probability density functions. This measure is decomposed into two parts, one a distance to the sieve and one a penalty on the sieve size. From this decomposition, an optimal rate of growth of the sieve is derived. In Section 6, some implementation issues including the computational complexity of the algorithm are discussed. A set of simulations is shown in Section 7.

2. Radar data model.

While the model in (1) is fairly general, our interest here is in the special case of narrowband radar imaging. Suppose that the transmitted radar signal has complex envelope [33] equal to $s_T(t)$. The complex-valued ratio of the signal reflected by a unit-area patch to the impinging signal is called the *reflectivity*. There are several assumptions which give rise to the received signal model. First reflections are assumed to be linear so that the definition of reflectivity makes sense. Second, the effects of creeping waves are ignored. Third, it is assumed that the bandwidth of $s_T(t)$ is small compared to the wavelength of the transmitted signal and that the pulse squeezing effects due to target motion are negligible. Then the received signal due to a reflection at delay τ and Doppler f is given by

$$(2) \qquad s_T(t-\tau)e^{j2\pi f(t-\tau/2)}z(f,\tau),$$

where $z(f,\tau)$ is the reflectivity. For a target moving at velocity v at a distance x, the Doppler-delay coordinates are given by (f_0 is the carrier frequency)

$$(3) \qquad (f,\tau) = \left(\frac{2vf_0}{c}, \frac{2x}{c}\right);$$

so τ is the two way delay to the target. For rotating rigid bodies, (3) indicates that there is a linear transformation taking Doppler-delay coordinates to crossrange-range coordinates.

The transmitted signal is assumed to be unfocused relative to the target dimensions (spotlight mode) so that the received signal is a superposition of the returns from patches across the entire surface of the target; the received signal is observed in the presence of an additive noise giving the model

$$(4a) \qquad r(t) = \int_{-f_{max}}^{f_{max}} \int_{0}^{\tau_{max}} s_T(t-\tau)e^{j2\pi f(t-\tau/2)}z(f,\tau)df\,d\tau + w(t).$$

The limits in (4a) are due to an assumption of finite target extent. The integral in (4a) may be rewritten as a Lebesgue-Stieltjes integral [25, §12.3]. Let $Z([f, f + df] \times [\tau, \tau + d\tau]) = Z(df, d\tau)$ be the total reflectivity of a patch located at $[f, f + df] \times [\tau, \tau + d\tau]$. Then, (4a) may be rewritten as

$$(4b) \qquad r(t) = \int_{-f_{max}}^{f_{max}} \int_{0}^{\tau_{max}} s_T(t-\tau)e^{j2\pi f(t-\tau/2)}Z(df, d\tau) + w(t).$$

2.1. Statistical assumptions. We adopt a diffuse target model. The target is assumed to have several independent scatterers in any resolution cell. This leads to a model of z as a zero mean, complex Gaussian random process with orthogonal random variables [12 §2.5], as discussed by Van Trees [33] and Snyder et al. [30]. This model assumes uncorrelated scatterers at different Doppler-delay coordinates. Thus the reflectivity process has independent increments and second order statistics given by

$$(5) \qquad E[Z([f, f+df] \times [\tau, \tau+d\tau])Z^*([f, f+df] \times [\tau, \tau+d\tau])] = S(f,\tau)df\,d\tau.$$

The function $S(f,\tau)$ is known as the *scattering function* of the target. This function represents the intensity of the reflectivity process in delay and Doppler coordinates and may be viewed as an image of the target. Since the reflectivity is a stochastic process, it is expected that the scattering function provides a better representation of the object than the reflectivity itself.

The additive noise $w(t)$ is modeled as zero mean complex-valued white Gaussian noise with intensity N_0

$$(6) \qquad E[w(t)w^*(t')] = N_0\delta(t - t').$$

Thus the autocovariance function for the random process $r(t)$ is

$$(7) \quad E[r(t)r^*(t')] = \int\limits_{-f_{max}}^{f_{max}} \int\limits_{0}^{\tau_{max}} s_T(t-\tau)s_T^*(t'-\tau)e^{j2\pi f(t-t')}S(f,\tau)df\,d\tau + N_0\delta(t-t').$$

There are several different formulations of the estimation problem. One may be stated as estimating the scattering function, $S(f,\tau)$, from a set of continuous time measurements, $r(t), t \in T$. This problem is difficult to solve directly as the likelihood function is a complicated functional of the scattering function. Another formulation is to estimate $S(f,\tau)$ given a set of samples of $r(t)$ or a set of samples of the output of a filter with $r(t)$ as input. The choice of the filter is important in this latter case. In either of these latter formulations, the estimation problem is ill-posed and the use of sieves is warranted.

For the problem of estimating the scattering function, the cross-ambiguity function of the received signal with the transmitted signal is a sufficient statistic (this is shown for a discrete version of the problem in (26)). This cross-ambiguity function is defined by

$$(8) \qquad \int\limits_{T} r(t)s_T^*(t-\tau)e^{-j2\pi ft}dt,$$

and may be formed by a set of matched filters, matched to the transmitted signal modulated by $\exp(j2\pi ft)$ for all f. There is a factor of $\exp(-j2\pi f\tau/2)$ which appears in (4), but neither in (8), nor in the covariance function (7). While this factor does not affect the likelihood function, it will be retained in our equations. Thus, we define the function

$$(9) \qquad q(f,\tau) = \int\limits_{T} r(t)s_T^*(t-\tau)e^{-j2\pi f(t-\tau/2)}dt$$

as our (slightly modified) cross-ambiguity function. Substituting (4) into (9) gives the cross-ambiguity function in terms of the ambiguity function of the transmitted signal as

$$
\begin{aligned}
(10) \qquad q(f,\tau) = E_T &\iint A_{ss}(f-f',\tau-\tau')e^{j\pi(f'\tau-f\tau')}z(f',\tau')df'd\tau' \\
&+ \int\limits_{T} w(t)s_T^*(t-\tau)e^{-j2\pi f(t-\tau/2)}dt,
\end{aligned}
$$

where the ambiguity function for $s_T(t)$ is defined by

$$(11) \qquad A_{ss}(f,\tau) = \frac{1}{E_T}\int s_T\left(t+\frac{\tau}{2}\right)s_T^*\left(t-\frac{\tau}{2}\right)e^{-j2\pi ft}dt.$$

Here $E_T = \int |s_T(t)|^2dt$ is the energy in $s_T(t)$.

Corresponding to the possible estimation problems outlined above, there are two scenarios for the collection of data. The first is a sampled version of $r(t)$. The second is a sampled version of $q(f, \tau)$. This paper is devoted to the first case. In the second case, the estimation algorithm relies on determining a likelihood for $q(f, \tau)$ in terms of the parameters in the sieve sets, then maximizing this likelihood function over those parameters. The principal change in the algorithms presented here is that the ambiguity matrix (to be defined below) is replaced by samples of the ambiguity function (11).

2.2. Discrete data. In this subsection, the discrete model is defined, and an analysis of the relationship of this model to the continuous model is presented. While the model is essentially the same as that used in [24,30], the mean-square convergence of this model to the continuous model was not addressed in those references. This is important in the following sections as the resulting estimate of the scattering function may be shown to converge to the true scattering function as the number of available data samples, N, goes to infinity and the number of parameters in the sieve tends to infinity at an appropriate rate.[1]

The available data are assumed to be samples, $r(n\Delta t)$, of the data $r(t)$ from (4). The contribution from the white noise component is assumed to be averaged over a small time period and $\{w(n\Delta t), 0 \leq n < N\}$ is modeled as a set of zero mean, independent, identically distributed, complex Gaussian random variables with variance N_0 (different from the N_0 in (6)). The covariance matrix for the discrete data has entries given by

(12)
$$K_r(n, m) = E[r(n\Delta t)r^*(m\Delta t)]$$

$$= \int_{-f_{\max}}^{f_{\max}} \int_0^{\tau_{\max}} s_T(n\Delta t - \tau)s_T^*(m\Delta t - \tau)e^{j2\pi f(n-m)\Delta t}S(f, \tau)df\,d\tau + N_0\delta_{nm},$$

$$0 \leq n, m < N.$$

Our numerical algorithms require a discretization of the scattering function. Let $\{\Pi_{ki}, 0 \leq k < I_{CR}, 0 \leq i < I_R\}$ be a partition of the region over which the scattering function is defined:

(13) $$\Pi_{ki} = [f_k, f_{k+1}] \times [\tau_i, \tau_{i+1}], \text{ where } f_{k+1} > f_k, \tau_{i+1} > \tau_i.$$

The subscripts CR and R refer to cross-range and range, respectively. Note that this partition is separable (it is the product of one-dimensional partitions). The size of cell Π_{ki} is

(14) $$|\Pi_{ki}| = (f_{k+1} - f_k)(\tau_{i+1} - \tau_i).$$

---

[1] The convergence of the scattering function is in terms of the discrepancy measure (Kullback-Leibler divergence) between the probability density functions corresponding to the true scattering function and corresponding to the estimate of the scattering function. If the true scattering function is bounded away from zero, then the estimate converges to the true scattering function in an L^1 sense [21].

Using this partition, a discrete approximation to the integral in (4) may be obtained. For each cell, let

$$c(k, i) = |\Pi_{ki}|^{-1/2} Z([f_k, f_{k+1}] \times [\tau_i, \tau_{i+1}]).$$

The Lebesgue-Stieltjes integral in (4b) may be approximated by the Riemann sum

(16)

$$\tilde{r}(n\Delta t) = \sum_{k=0}^{I_{CR}-1} \sum_{i=0}^{I_R-1} |\Pi_{ki}|^{1/2} e^{j2\pi f_k(n\Delta t - \tau_i/2)} s_T(n\Delta t - \tau_i) c(k, i) + w(n\Delta t), 0 \le n < N.$$

By definition of the stochastic integral in (4b), the sum in (16) converges to that integral in the mean-square sense. That is,

(17)
$$\lim E[|r(n\Delta t) - \tilde{r}(n\Delta t)|^2] = 0,$$

where the limit is taken as the maximum cell-size, $\max_{i,k} |\Pi_{ki}|$, goes to zero. Thus, the discrete model for the reflectivity is asymptotically equivalent to the continuous model in the sense of (17).

The $c(k, i)$ are put into the $(k + iI_{CR})$ entry of an $I_{CR}I_R \times 1$ vector c. A matrix $\Gamma^\dagger$ is defined to have $(n, k + iI_{CR})$ entry

(18)
$$|\Pi_{ki}|^{1/2} e^{j2\pi f_k(n\Delta t - \tau_i/2)} s_T(n\Delta t - \tau_i),$$

where the superscript $\dagger$ denotes Hermitian transpose. Equation (16) is used as a model for the actual data $r(n\Delta t)$. Placing the $r(n\Delta t)$ and $w(n\Delta t)$ in vectors r and w, respectively gives the discrete data model

(19)
$$r = \Gamma^\dagger c + w.$$

The probability density function for r subject to the discrete model (19) is parameterized by a discrete approximation of the scattering function. From the model, the entries of the c vector are zero mean, independent, complex Gaussian random variables with variances equal to

(20)
$$E[|c(k, i)|^2] = \sigma^2(k, i) = \frac{1}{|\Pi_{ki}|} \int_{f_k}^{f_{k+1}} \int_{\tau_i}^{\tau_{i+1}} S(f, \tau) df d\tau.$$

The covariance matrix for r for the discrete model is then given by

(21)
$$K_r = E[rr^\dagger] = \Gamma^\dagger \Sigma \Gamma + N_0 I_N,$$

where Σ is an $I_{CR}I_R \times I_{CR}I_R$ diagonal matrix with diagonal entries $\sigma^2(k, i)$, and I_N is the $N \times N$ identity matrix. The entries of K_r converge to $K_r(n, m)$ from (12) as the maximum cell-size goes to zero.

2.3. Likelihood function. The likelihood function for r depends on the model adopted. In the estimation problem, the likelihood is parameterized by either the scattering function or a discrete version of the scattering function. Then the likelihood is maximized subject to that parameterization. For the continuous model, the data $r(n\Delta t)$ have covariance matrix K_r from (12) and the loglikelihood function is

$$(22) \qquad l(r; S) = -\ln \det(K_r) - r^\dagger (K_r)^{-1} r.$$

Rather than directly attempting to maximize $l(r; S)$ over all nonnegative functions $S(f, \tau)$, a two step approximation is taken. The first step is to assume the discrete model (19). The loglikelihood for (19) is

$$(23) \qquad l(r; \Sigma) = -\ln \det(\Gamma^\dagger \Sigma \Gamma + N_0 I_N) - r^\dagger (\Gamma^\dagger \Sigma \Gamma + N_0 I_N)^{-1} r.$$

The second step is to assume that Σ is in a sieve set. The ML estimate subject to a sieve constraint is that scattering function in the sieve set which maximizes (23).

The likelihood function may be rewritten using the matrix identity

$$(24) \qquad (\Gamma^\dagger \Sigma \Gamma + N_0 I_N)^{-1} = \frac{1}{N_0} I_N - \frac{1}{N_0} \Gamma^\dagger (\Sigma \Gamma \Gamma^\dagger + N_0 I_N)^{-1} \Sigma \Gamma$$

as

$$(25) \quad l(r; \Sigma) = -\ln \det(\Gamma^\dagger \Sigma \Gamma + N_0 I_N) - \frac{1}{N_0} r^\dagger r + \frac{1}{N_0} r^\dagger \Gamma^\dagger (\Sigma \Gamma \Gamma^\dagger + N_0 I_N)^{-1} \Sigma \Gamma r.$$

Thus, a sufficient statistic for the discrete problem is the discrete cross-ambiguity function

$$(26) \qquad q = \Gamma r.$$

The matrix

$$(27) \qquad A = \Gamma \Gamma^\dagger,$$

which appears in (25) (and implicitly in (26) multiplying c) is the discrete ambiguity matrix. The ambiguity matrix plays the role in the discrete problem that the ambiguity function in (10) and (11) plays for the continuous problem.

3. Conventional imaging. There are many alternative methods for forming images in radar problems. Many of these methods may be described by linear processing of the data, and many may be categorized as one of two possibilities.

Classical imaging techniques form the cross-ambiguity function and display its magnitude or magnitude squared as an image of the target. For these approaches, the entries of the vector q from (26) (or its entries magnitude squared) are displayed in delay-Doppler coordinates. For the magnitude squared image, the expected values of the display are the diagonal entries of the matrix

$$(28) \qquad A\Sigma A + N_0 A.$$

The $N_0 A$ term is due to the noise. Clearly the expectation of the image is a blurred and biased version of the desired image (Σ).

Other image formation algorithms attempt to invert the effect of the transmitted signal to recover the vector c directly. Such approaches are motivated by a desire to avoid the blurring in (28). The algorithms may be characterized by multiplying the vector r by the inverse of the matrix $\Gamma^\dagger$ (assuming its invertibility):

$$(29) \qquad p = (\Gamma^\dagger)^{-1} r.$$

Then the magnitude or magnitude squared of the entries of p are displayed as an image. For the magnitude squared display, the expected values are the diagonal entries of the matrix

$$(30) \qquad \Sigma + N_0 A^{-1}.$$

While the effect of the noise term, $N_0 A^{-1}$, may be partially removed by thresholding the display, the resulting image will still be very rough because the display consists of exponentially-distributed random variables. For $N_0 = 0$, the display above is also the direct ML solution to (23). The estimates have high variance because the number of parameters relative to the number of data in the estimation problem is too large.

Thus, the need for regularization is motivated by the roughness in displays of high-variance statistics. The sieve method proposed in the next section aims at exploiting the well-known asymptotic properties of maximum likelihood estimators.

4. Method of sieves.

Stable solutions to ill-posed estimation problems can be obtained using the *method of sieves* due to Grenander [13]. Here a restricted set of possible solutions is defined, and the likelihood functional is maximized subject to the constraint that the parameter be a member of that set.

Formally, a *sieve in a parameter space* $\mathbf{A}$ is a family of subsets $\mathbf{S}(\mu)$ of $\mathbf{A}$ indexed by a positive parameter μ called the *mesh size* [13 §7.1]. A restricted ML estimate exists over each set $\mathbf{S}(\mu)$. As the mesh size tends to zero, the sets $\mathbf{S}(\mu)$ will be large enough to allow the ML solution to converge to any solution in $\mathbf{A}$.

A major problem is to find sieves which will make the ML estimates converge and possess some desirable practical features such as analytical and/or computational tractability. Typically μ is some measure of smoothness of the functions in a sieve subset and is chosen according to the data record size, the noise level, and possibly other factors. Sieves have been used in various areas, including probability density estimation [13] and medical imaging [19,28,29]. Often, finding estimates within a sieve presents severe computational problems.

Under certain conditions, sieves can be designed to ensure consistency of the estimates. Generally speaking, this requires that the mesh size μ of the sieve decreases as the number of observations N increases. This decrease should be "slow enough". In this fashion, the sieve subsets are large enough to ensure a good representation of any function in the parameter set $\mathbf{A}$, yet they are also small enough to guarantee stable estimates. Two challenging problems are determining what rates of decrease lead to consistent estimates and finding optimal rates.

The method of sieves recommended here is based on a spline representation for the unknown scattering function. Such representations have been used in statistics for ML probability density estimation [4]. Splines have also been proposed for the related problem of power-density estimation. Cogburn and Davis have studied a spline smoothing technique for the empirical spectrum estimator (periodogram) [9]. This technique yields estimates in a class of splines, albeit not ML estimates within this class. Our approach differs in the sense that we are seeking ML estimates.

In Sections 4.1 and 4.2, we show how to obtain ML estimates in a class of functions defined by linear combinations of known elementary functions. In Section 4.3, we apply the method to the special case of splines.

4.1. Sieve. We denote by $\Omega := [-f_{\max}, f_{\max}] \times [0, \tau_{\max}]$ the domain of the scattering function. The scattering function itself belongs to some linear space $\mathbf{B}$ of real-valued functions over Ω. Since S is nonnegative, it also belongs to $\mathbf{B}_+$, the subset of all nonnegative functions in $\mathbf{B}$. Typically $\mathbf{B}$ can be $\mathbf{L}^1(\Omega)$, the space of absolutely integrable functions over Ω. The $\mathbf{L}^1$ norm of S must be finite, since this quantity is equal to the total expected energy in the reflectivity process. In other problems, $\mathbf{B}$ could be a more restricted space.

Define a countable index set Λ, and let $\{\psi_m(f, \tau), m \in \Lambda\}$ be a set of (nonnegative) elementary functions in $\mathbf{B}_+$. It is not required that this set be orthogonal. Consider the set $\mathbf{A}$ of all functions in the span of $\{\psi_m(f, \tau), m \in \Lambda\}$, with nonnegative coefficients:

$$(31) \qquad \mathbf{A} := \left\{ S \in \mathbf{B} \,\middle|\, S(f, \tau) = \sum_{m \in \Lambda} a(m)\psi_m(f, \tau), a \geq 0 \right\}$$

We define our parameter set to be $\mathbf{A}$. The functions in this set are represented by a countable number of coefficients $\{a(m), m \in \Lambda\}$. Clearly $\mathbf{A}$ is a subset of $\mathbf{B}_+$. A requirement is that $\{\psi_m(f, \tau), m \in \Lambda\}$ be designed so that any function in $\mathbf{B}_+$ can be approximated with arbitrary accuracy by a function in $\mathbf{A}$ under an appropriate "distance". We define a sieve $\mathbf{S}(\Lambda_Q)$ on S in $\mathbf{A}$ by truncating the representation (31) to a finite number of terms Q,

$$(32) \qquad \mathbf{S}(\Lambda_Q) := \left\{ S \in \mathbf{B} \,\middle|\, S(f, \tau) = \sum_{m \in \Lambda_Q} a(m)\psi_m(f, \tau), a \geq 0 \right\},$$

where the index set Λ_Q has cardinality Q, and the union of all sets $\Lambda_Q, Q \in \mathbf{N}$ is equal to Λ. The functions in the subset $\mathbf{S}(\Lambda_Q)$ of $\mathbf{A}$ are represented by means of a finite number Q of coefficients $a(m), m \in \Lambda_Q$. We view $\{a(m), m \in \Lambda_Q\}$ as a set of Q unknown parameters for which ML estimates are sought. In the next section, we show how estimates can be produced numerically using an Expectation-Maximization (EM) algorithm.

For the discrete model (19), a sieve can be defined in a similar manner. Here the space of scattering functions $\mathbf{B}$ has dimension $I_{CR} I_R$. Each elementary function is represented as a vector $\psi_m^{(I_{CR},I_R)} = \{\psi_m^{(I_{CR},I_R)}(k,i) = \psi_m(f_k,\tau_i), 0 \leq k < I_{CR}, 0 \leq i < I_R\}$. The discrete parameter set is defined by

$$(33) \quad \mathbf{A}^{(I_{CR},I_R)} := \left\{ \sigma^2 \in \mathbf{B} | \sigma^2(k,i) = \sum_{m \in \Lambda} a(m)\psi_m^{(I_{CR},I_R)}(k,i), a \geq 0 \right\}.$$

We define a sieve $\mathbf{S}^{(I_{CR},I_R)}(\Lambda_Q)$ in $\mathbf{A}^{(I_{CR},I_R)}$ by

$$(34) \quad \mathbf{S}^{(I_{CR},I_R)}(\Lambda_Q) = \left\{ \sigma^2 \in \mathbf{B} | \sigma^2(k,i) = \sum_{m \in \Lambda_Q} a(m)\psi_m^{(I_{CR},I_R)}(k,i), a \geq 0 \right\},$$

where the integer Q is constrained to be no larger than $I_{CR}I_R$.

4.2. ML estimator. In this section, the maximum likelihood problem is defined, conditions for a maximizer are derived, and our algorithm for obtaining the maximum likelihood estimate is presented. These results extend those in [24,30] by enforcing the sieve constraint and reducing the computational complexity of the solution.

4.2.1. Likelihood function. The loglikelihood function for the parameters a is given by

$$(35) \quad l(r;a) = -\ln \det K_r - r^\dagger K_r^{-1} r.$$

From (21) and (32), the dependence of K_r on a is as follows,

$$(36) \quad K_r = \sum_{m \in \Lambda_Q} a(m)K_m + N_0 I_N,$$

where we have defined the $N \times N$ *basis covariance matrix*

$$(37) \quad K_m = \left(\int_\Omega s_T(n\Delta t - \tau)s_T^*(p\Delta t - \tau)e^{j2\pi f(n-p)\Delta t}\psi_m(f,\tau)df d\tau \right)_{0 \leq n,p < N}.$$

The maximization of (35), subject to the constraint (36), yields sieve-constrained ML estimates of the scattering function. As shown by (35) and (36), the structure of the new estimation problem is determined by the set of basis covariance matrices $\{K_m\}$. Anderson [3] studied a similar problem of covariance matrix estimation in which the unknown covariance matrix is a linear combination of given matrices. This problem can be viewed as a special case of ours where it is known that the covariance is in a given subset of the sieve. Anderson derived expressions for the gradient and the Hessian of the likelihood function in the case of real matrices. He also proposed a Newton-Raphson method for numerically finding the maximizer of the likelihood function.

For the discrete model (19), the loglikelihood is still given by (35) and (36), in which K_m is a discrete version of (37). Introducing the diagonal $I_{CR}I_R \times I_{CR}I_R$ matrix $\Psi_m = \text{diag}\{\psi_m(f_k, \tau_i)\}_{0 \leq k < I_{CR}, 0 \leq i < I_R}$, we have

$$(38) \qquad K_m = \Gamma^\dagger \Psi_m \Gamma,$$

The continuous and discrete estimation problems differ only by the expressions for the basis covariances, given by (37) and (38), respectively. For fixed N, K_m in (37) is the limit of (38) as $I_{CR}, I_R \to \infty$. It is thus equivalent to maximize the loglikelihood for the continuous or the discrete model, in the limit as $I_{CR}, I_R \to \infty$. These results are consistent with the asymptotic equivalence of the two models established in Section 2.2.

4.2.2. Conditions for a maximizer. The loglikelihood (35) is to be maximized subject to nonnegativity constraints on the components of a. The necessary conditions for some $\hat{a}$ to be a local maximum of the loglikelihood can be derived from the Kuhn-Tucker conditions [18]. In order to establish these conditions, we first derive the gradient and the Hessian matrix of the loglikelihood function. Related work includes [7,30].

For $K(x)$ be a square matrix parameterized by some real x, we have the equalities $\frac{dK^{-1}}{dx} = -K^{-1}\frac{dK}{dx}K^{-1}$ and $\frac{d\ln\det K}{dx} = Tr\left[\frac{dK}{dx}K^{-1}\right]$. By application of these equalities and (36), the mth component of the gradient of the loglikelihood (35) is

$$(39) \qquad \frac{\partial l(r; a)}{\partial a(m)} = Tr[K_m K_r^{-1}(rr^\dagger - K_r)K_r^{-1}].$$

The (m, n) entry of the Hessian matrix is given by

$$(40) \qquad H_{mn} = -\text{ Re } Tr[K_m K_r^{-1} K_n K_r^{-1}(2rr^\dagger - K_r)K_r^{-1}].$$

The necessary and sufficient conditions for a maximizer of the loglikelihood are contained in the following two propositions.

PROPOSITION 1. *A necessary condition for K_r to be a local maximum of the loglikelihood is*

$$(41) \qquad Tr[K_m K_r^{-1}(rr^\dagger - K_r)K_r^{-1}] + \mu(m) = 0, \qquad m \in \Lambda_Q,$$

where $\mu(m) = 0$ if $a(m) > 0$, and $\mu(m) \geq 0$ if $a(m) = 0$.

Proof. Equation (41) is just the necessary Kuhn-Tucker condition for maximizing a function subject to the inequality constraints $\{a(m) \geq 0, m \in \Lambda_Q\}$ [18]. The first term in the left-hand side of (41) is the gradient of the objective function. Each Lagrange multiplier $\mu(m)$ is zero if the corresponding constraint is inactive, and negative otherwise. Note that condition (41) expresses that for every admissible variation δa of the parameter (i.e. $a + \delta a \geq 0$), the variation of the likelihood is nonpositive:

$$\sum_{m \in \Lambda_Q} \delta a(m) Tr[K_m K_r^{-1}(rr^\dagger - K_r)K_r^{-1}] \leq 0. \quad \square$$

In the following, we refer to (41) as the *trace condition*.

PROPOSITION 2. *A sufficient condition for K_r to be a local maximum of the loglikelihood is that*

(1) *The trace condition (41) is satisfied.*

(2) *The Hessian matrix (40) is negative definite.*

Proof. Follows from a straightforward application of the Kuhn-Tucker conditions. □

4.2.3. Closed-form solutions. The trace condition (41) is a nonlinear equation in the Q unknown parameters a. In general, it cannot be solved in closed form, and some numerical method such as the EM algorithm below must be used. However, we mention a special instance of the discrete model for which closed-form solutions can be derived. Let $I_{CR}I_R = N$, Γ be invertible, and assume that either $N_0 = 0$ or that the ambiguity matrix (27) equals $\epsilon_T I_N$. Note that ϵ_T is a discrete approximation to the transmitted signal energy E_T. These conditions are satisfied, for instance, when a stepped-frequency waveform is used as a transmitted signal and the cells Π_{ki} are all the same size. We study the particular elementary functions $\{\psi_m^{(I_{CR},I_R)}(k,i), m \in \Lambda_Q\}$ which are indicator functions over disjoint regions $\{\mathbf{D}_m, m \in \Lambda_Q\}$. With the assumptions on Γ, we have

$$K_r = \Gamma^\dagger \Sigma \Gamma + N_0 I_N = \Gamma^\dagger (\Sigma + N_0/\epsilon_T I_N)\Gamma.$$

The trace condition (41) takes the following form:

$$0 = Tr[\Gamma^\dagger \Psi_m \Gamma(\Gamma^\dagger \Sigma \Gamma + N_0/\epsilon_T I_N)^{-1}(rr^\dagger - \Gamma^\dagger \Sigma \Gamma - N_0/\epsilon_T I_N)(\Gamma^\dagger \Sigma \Gamma + N_0/\epsilon_T I_N)^{-1}] + \mu(m)$$

$$= Tr[\Psi_m(\Sigma + N_0/\epsilon_T I_N)^{-1}(pp^\dagger - \Sigma - N_0/\epsilon_T I_N)^{-1})(\Sigma + N_0/\epsilon_T I_N)^{-1}] + \mu(m)$$

$$= \sum_{k,i \in \mathbf{D}_m} \psi_m^{(I_{CR},I_R)}(k,i)\left[\frac{|p(k,i)|^2}{(\sigma^2(k,i) + N_0/\epsilon_T)^2} - \frac{1}{\sigma^2(k,i) + N_0/\epsilon_T}\right] + \mu(m), m \in \Lambda_Q,$$

where p is the preimage defined in (29). Since the $\{\psi_m^{(I_{CR},I_R)}\}$ are indicator functions, the right-hand side above is equal to

$$\frac{\sum\limits_{k,i \in \mathbf{D}_m} |p(k,i)|^2}{(a(m) + N_0/\epsilon_T)^2} - \frac{|\mathbf{D}_m|}{a(m) + N_0/\epsilon_T} + \mu(m)$$

If the constraint is inactive, $\mu(m) = 0$, the parameter $a(m)$ is nonnegative and is given by

$$a(m) = \frac{1}{|\mathbf{D}_m|} \sum_{k,i \in \mathbf{D}_m} |p(k,i)|^2 - N_0/\epsilon_T.$$

On the other hand, if the constraint is active, we have necessarily $a(m) = 0$. The solution can be written in the more condensed form

$$a(m) = \max\left\{\frac{1}{|\mathbf{D}_m|} \sum_{k,i \in \mathbf{D}_m} |p(k,i)|^2 - N_0/\epsilon_T, 0\right\}, \quad m \in \Lambda_Q.$$

4.2.4. EM algorithm. The EM algorithm of Dempster, Laird and Rubin [11] is based on the concept of complete and incomplete data spaces and the existence of a many-to-one mapping from the former space to the latter. At each step of the algorithm, the conditional expectation of the complete-data likelihood is evaluated, where the conditioning is on the observations and on the current parameter estimate. Next, this expectation is maximized with respect to the parameter. The maximizer is an updated estimate. Dempster et al. have shown that this alternating maximization algorithm yields a sequence of estimates having nondecreasing likelihood.

We propose the following EM algorithm for computing the estimates. Write the discrete reflectivity process c as the sum of Q independent processes:

$$(42) \qquad c = \sum_{m \in \Lambda_Q} c_m,$$

where c_m is a $I_{CR}I_R$-vector, sample of a 0-mean Gaussian process with diagonal covariance $a(m)\Psi_m$. The support set of the elementary function $\psi_m^{(I_{CR}, I_R)}(k, i)$ is denoted by $\mathbf{D}_m$. With (42), the model (19) becomes

$$(43) \qquad r = \sum_{m \in \Lambda_Q} \Gamma^\dagger c_m + w.$$

We define the complete data as $(\{c_m, m \in \Lambda_Q\}; w)$. Because the Q processes $\{c_m\}$ are independent, the loglikelihood for these data is simply

$$
\begin{aligned}
l_{cd}(a) : &= \ln p(c : a) \\
&= \sum_{m \in \Lambda_Q} \ln p(c_m : a) \\
&= - \sum_{m \in \Lambda_Q} \ln \det(a(m)\Psi_m) - \sum_{m \in \Lambda_Q} c_m^\dagger (a(m)\Psi_m)^{-1} c_m.
\end{aligned}
$$

Discarding all terms not involving a, we obtain:

$$
l_{cd}(a) = - \sum_{m \in \Lambda_Q} |\mathbf{D}_m| \ln a(m) - \sum_{m \in \Lambda_Q} \frac{1}{a(m)} \sum_{k,i \in \mathbf{D}_m} \frac{|c_m(k,i)|^2}{\psi_m^{(I_{CR}, I_R)}(k, i)}.
$$

Maximizing the conditional expectation of $l_{cd}(a)$ with respect to $a(m)$ at step p of the algorithm, we obtain the updated estimate for the following step:

$$(44) \qquad \hat{a}(m)^{(p+1)} = \frac{1}{|\mathbf{D}_m|} \sum_{k,i \in \mathbf{D}_m} \frac{E[|c_m(k,i)|^2 | r, \hat{a}^{(p)}]}{\psi_m^{(I_{CR}, I_R)}(k, i)}.$$

Next we find the conditional expectation of $|c_m(k,i)|^2$ in (44). Define $\hat{c}_m^{(p)} := E[c_m | r, \hat{a}^{(p)}]$. The conditional expectation in (44) is the $(k + iI_{CR})^{\text{th}}$ diagonal element of the matrix

$$
\begin{aligned}
E[c_m c_m^\dagger | r, \hat{a}^{(p)}] &= E[(c_m - \hat{c}_m^{(p)})(c_m - \hat{c}_m^{(p)})^\dagger | r, \hat{a}^{(p)}] + \hat{c}_m^{(p)} \hat{c}_m^{(p)\dagger} \\
&= K_{c_m c_m}^{(p)} - K_{c_m r}^{(p)} K_r^{(p)-1} K_{r c_m}^{(p)} + (K_{c_m r}^{(p)} K_r^{(p)-1} r)(K_{c_m r}^{(p)} K_r^{(p)-1} r)^\dagger,
\end{aligned}
$$

where we denote by K_{xy} the conditional correlation $E[xy^\dagger|\hat{a}^{(p)}]$ of two random vectors x and y. Now, the expectations are evaluated from the model (43), taking into account the independence of the c_m's. We obtain

$$K^{(p)}_{c_m r} = E[c_m r^\dagger|\hat{a}^{(p)}] = \hat{a}(m)^{(p)}\Psi_m\Gamma,$$
$$K^{(p)}_r = E[rr^\dagger|\hat{a}^{(p)}] = \sum_{j\in\Lambda_Q} \hat{a}(j)^{(p)}\Gamma^\dagger_j\Psi_j\Gamma_j + N_0 I_N,$$
$$K^{(p)}_{c_m c_m} = E[c_m c^\dagger_m|\hat{a}^{(p)}] = \hat{a}(m)^{(p)}\Psi_m.$$

Substituting these expressions and simplifying, we obtain the update equations at stage p of the algorithm:

$$(45) \qquad\qquad K^{(p)}_r = \sum_{j\in\Lambda_Q} \hat{a}(j)^{(p)}K_j + N_0 I_N.$$

$$(46) \qquad\qquad Z^{(p)} = K^{(p)-1}_r(rr^\dagger - K^{(p)}_r)K^{(p)-1}_r,$$

$$(47) \qquad \hat{a}(m)^{(p+1)} = \hat{a}(m)^{(p)} + \frac{1}{|\mathbf{D}_m|}\left(\hat{a}(m)^{(p)}\right)^2 Tr\left[K_m Z^{(p)}\right], \quad m\in\Lambda_Q.$$

Comments.

(1) As appears from (44), the estimates are guaranteed to be nonnegative.

(2) The trace appearing in the update equation (47) is the mth component of the gradient of the loglikelihood in (39).

(3) The stable points of the algorithm satisfy the necessary Kuhn-Tucker conditions of Proposition 1. This follows directly from (47) and comment (2) above.

(4) Computing the update term in (47) involves only "time-domain" equations. These equations highlight the central role played by the basis covariances K_m in this maximization problem.

(5) Each matrix K_m is constant and can be computed off-line. At each step the covariance matrix $K^{(p)}_r$ must be inverted, which accounts for N^3 operations. Following this, updating each $\hat{a}(m)$ requires N^2 multiplications/additions. The complexity of each step of the EM algorithm is then $N^3 + QN^2$.

(6) Increasing $I_{CR}I_R$ has two effects:

· The convergence of the EM algorithm is slowed down, due to the $1/|\mathbf{D}_m|$ term in the update equation (47). This is not a severe problem.

· As mentioned before, the basis covariance matrix (38) converges to (37). Thus, $I_{CR}I_R$ can be made arbitrarily large without any loss in stability of the estimates or any major increase in computations.

4.3. Spline sieve. For the representation of functions in the sieve, we select a set of polynomial splines as elementary functions. This permits a good *local* representation of a function. This approach offers several advantages. First, the local behavior of the function depends only on a few parameters. This property is useful in the context of imaging. In contrast, with Fourier and other polynomial representations, the local behavior of the function depends on all Fourier coefficients. Locality also leads to a considerable reduction in the complexity of the estimation algorithm, as discussed in Section 6. Another advantage of the method is that the resolution and smoothness of the splines can be controlled and their approximation power evaluated.

Consider an interval $[a, b]$ of the real line, over which a partition $\{x_m, 0 \le m < M\}$ is defined, with $a = x_0 < x_1 < \cdots < x_{M-1} = b$. This partition is not necessarily uniform. Let $f(x)$ be a function over $[a, b]$ so that $f(x)$ agrees with a polynomial of degree $L-1$ over each interval $[x_m, x_{m+1}]$, and the $L-2$ first derivatives of $f(x)$ are continuous at the knot points. By definition, $f(x)$ is a polynomial spline of order L with simple knots at the points $x_0, \ldots x_{M-1}$. The smoothness of $f(x)$ is determined by the order L of the spline. The resolution level at some $x \in [a, b]$ is the size of the partition cell to which x belongs.

In the following, we use Schoenberg's B-splines [26], since they possess attractive mathematical properties. The partition $\{x_m, 0 \le m < M\}$ is extended on each side of $[a, b]$ with $L - 1$ knots chosen arbitrarily, so long as $x_{-L+1} \le \cdots \le x_0 = a$ and $b = x_{M-1} \le \cdots \le x_{M+L-2}$. We denote by $\{B_m^{(L)}(x), -L + 1 \le m < M\}$ the normalized B-splines associated with the extended partition. The support set of $B_m^{(L)}(x)$ is $[x_m, x_{m+L}]$. In the special case of a uniform partition, $B_m^{(L)}(x)$ has the form $B^{(L)}(M(x - a)/(b - a) - m)$, where $B^{(L)}(x)$ is the basic B-spline, an $(L - 1)$-fold convolution of the characteristic function of $[0, 1]$ with itself. For a complete discussion of normalized B-splines, see for instance [27 §4.3].

The extension to two dimensions is carried in the simplest possible manner, using tensor-product concepts. The spline functions are then piecewise tensor-product (i.e. separable) polynomials with knots on a two-dimensional grid. They are constructed on Ω as follows. Define extended one-dimensional partitions $\mathbf{P}_R := \{\tilde{\tau}_{m_R}, -L+1 \le m_R < M_R+L-1\}$ and $\mathbf{P}_{CR} := \{\tilde{f}_{m_{CR}}, -L+1 \le m_{CR} < M_{CR}+L-1\}$ in range and cross-range coordinates respectively. Denote by $\{B_{R,m_R}^{(L)}(\tau), -L + 1 \le m_R < M_R\}$ and $\{B_{CR,m_{CR}}^{(L)}(f), -L + 1 \le m_{CR} < M_{CR}\}$ the normalized B-splines associated with the partitions $\mathbf{P}_R$ and $\mathbf{P}_{CR}$ respectively, and by $\mathbf{P}$ the tensor product $\mathbf{P}_{CR} \times \mathbf{P}_R$. Thus, $\mathbf{P}$ is a separable partition and can be written in the form

$$\mathbf{P} := \{\mathbf{P}_{m_{CR}, m_R} = [\tilde{f}_{m_{CR}}, \tilde{f}_{m_{CR}+L}] \times [\tilde{\tau}_{m_R}, \tilde{\tau}_{m_R+L}],$$
$$-L + 1 \le m_{CR} < M_{CR}, -L + 1 \le m_R < M_R\}.$$

The tensor-product B-splines defined over $\mathbf{P}$,

$$B_{m_{CR}, m_R}^{(L)}(f, \tau) = B_{CR, m_{CR}}^{(L)}(f) B_{R, m_R}^{(L)}(\tau), \quad -L+1 \le m_{CR} < M_{CR}, -L+1 \le m_R < M_R,$$

have support set $[\tilde{f}_{m_{CR}}, \tilde{f}_{m_{CR}+L}] \times [\tilde{\tau}_{m_R}, \tilde{\tau}_{m_R+L}]$. The *overlapping factor* ρ, defined as the number of elementary functions intersecting at any given point, is equal to L^2. For $L = 1$, $\{B_{m_{CR}, m_R}^{(L)}(f, \tau)\}$ are nonoverlapping "boxes"; for $L = 2$, $\{B_{m_{CR}, m_R}^{(L)}(f, \tau)\}$ are pyramids and for $L = 3$, $\{B_{m_{CR}, m_R}^{(L)}(f, \tau)\}$ are biquadratic functions. For a uniform partition, the splines are translates of the basic tensor-product spline $B^{(L)}(f)B^{(L)}(\tau)$ contracted at resolution levels $M_R/\tau_{\max}$ and $M_{CR}/(2f_{\max})$ in range and cross-range respectively,

$$B^{(L)}_{m_{CR},m_R}(f,\tau) = B^{(L)}(M_{CR}(f + f_{\max})/(2f_{\max}) - m_{CR})B^{(L)}(M_R\tau/\tau_{\max} - m_R).$$

We define a sieve based on splines as follows. The functions in the sieve are tensor-product polynomial spline functions. The degree $L - 1$ of the polynomial approximation is chosen according to the desired smoothness of the resulting scattering function. Each subset in the sieve is indexed by the partition that supports the spline. The mesh size of the sieve can be defined as the mesh size of the partition, that is, the size of its largest cell. As the partition is made finer, the size of the sieve subset increases in the space of functions S. The elementary functions are the tensor-product B-splines defined over $\mathbf{P}$. The number of elementary functions is $Q := (M_{CR} + L - 1)(M_R + L - 1)$.

In general, the sieve subsets are defined as

(48)
$$\mathbf{S}^{(L)}_{\mathbf{P}} := \left\{ S \in \mathbf{B} \mid S(f,\tau) = \sum_{m_{CR}=-L+1}^{M_{CR}-1} \sum_{m_R=-L+1}^{M_R-1} a(m_{CR},m_R)B^{(L)}_{m_{CR},m_R}(f,\tau), a \geq 0 \right\}.$$

For a uniform partition, (48) takes the form

(49)
$$\mathbf{S}^{(L)}_{M_{CR},M_R} := \left\{ S \in \mathbf{B} \mid S(f,\tau) = \sum_{m_{CR}=-L+1}^{M_{CR}-1} \sum_{m_R=-L+1}^{M_R-1} a(m_{CR},m_R) \right.$$
$$\left. \times B^{(L)}(M_{CR}(f + f_{\max})/(2f_{\max}) - m_{CR})B^{(L)}(M_R\tau/\tau_{\max} - m_R), a \geq 0 \right\}.$$

In this instance, the subsets of the sieve are simply indexed by M_{CR} and M_R, and the mesh size is $\|\mathbf{P}\| = |\mathbf{\Omega}|/(M_{CR}M_R)$.

It can be shown that the family of spline subsets (48) satisfies certain conditions for a sieve [21]. In particular:

(C1) There exists a ML estimate over each restricted set, almost surely, and

(C2) Any function in the parameter set can be approximated with arbitrary accuracy by a function in the sieve under the "distance" defined below, for $\|\mathbf{P}\|$ small enough.

5. Rate of growth of the sieve.

In this section, we study how the sieve should be designed to ensure consistency of the estimates. The issues are determining what rates of growth lead to consistent estimates and finding optimal rates. Convergence and consistency are measured with respect to an appropriate information distance. The mesh size of the sieve is then selected by minimizing the estimation error measured in this fashion.

5.1. Measure of the estimation error. The definition of the estimation error measure is first given for a general parameter estimation problem and is then specialized to the radar imaging problem. Consider two probability density functions $f_{\theta_0}(x)$ and $f_{\theta_1}(x)$ parameterized by θ_0 and $\theta_1 \in \mathbf{A}$, $x \in \mathbf{R}^N$. A measure of the "distance" between θ_0 and θ_1 is given by the average Kullback-Leibler information per sample:

$$\bar{d}^{(N)}(\theta_0 : \theta_1) := (1/N)I(f_{\theta_0} : f_{\theta_1})$$

(50)
$$= (1/N) \int f_{\theta_0}(x) \ln \frac{f_{\theta_0}(x)}{f_{\theta_1}(x)} dx.$$

We also assume that (50) converges to a limit $\bar{d}(\theta_0 : \theta_1)$ at the rate $O(N^{-1})$, as N tends to infinity. The functional (50) is positive, nonsymmetric in its arguments, and does not satisfy the triangle inequality. It is thus not a distance. Functionals of this type are sometimes called *directed distances* [16]. In the case where θ_1 is a function of a random variable x, a sample from the distribution $f_{\theta_0}(x)$, the left-hand side of (50) is also a random variable. We would like to define a measure which is independent of x. This is done by extending the definition of the distance to

(51) $\quad d^{(N)}(\theta_0 : \theta_1) := E_x[\bar{d}^{(N)}(\theta_0 : \theta_1(x))] = \int f_{\theta_0}(x)(1/N)I(f_{\theta_0} : f_{\theta_1(x)})dx.$

Clearly, when θ_1 is independent of x, the definitions (50) and (51) are equivalent. The definition (51) can be applied to an estimation problem in which $\hat{\theta}(x)$ is an estimate for the unknown parameter θ, based on the observation x. The estimation error is defined as

(52) $$d^{(N)}(\theta : \hat{\theta}).$$

The definition of the estimation error measure (51) is first applied to a general sieve problem. The subsets of the sieve are parameterized by a Q-vector parameter, as is the case for the spline sieve; we denote these subsets by $\mathbf{S}_Q$. Under some mild regularity conditions [14], as $N \to \infty$, there exists an asymptotic ML estimator $\tilde{\theta}$ within each $\mathbf{S}_Q$, for fixed μ. The following two lemmas, which are derived from the properties of the directed distance (51), are powerful tools for analysis of the error.

LEMMA 1. (Error decomposition).

(53) $$d^{(N)}(\theta : \hat{\theta}) \sim \bar{d}(\theta : \tilde{\theta}) + \frac{Q}{2N} \quad \text{as } N \to \infty.$$

where the notation $f(x) \sim g(x)$ as $x \to x_0$ means that f is asymptotic to g as $x \to x_0$, i.e. $\lim_{x \to x_0} \frac{f(x)}{g(x)} = 1$.

Proof.

$$d^{(N)}(\theta : \hat{\theta}) = (1/N)E_r \left[\int f_\theta(x) \ln \frac{f_\theta(x)}{f_{\hat{\theta}(r)}(x)} dx \right]$$

$$= (1/N)E_r \left[\int f_\theta(x) \ln \frac{f_\theta(x)}{f_{\tilde{\theta}}(x)} dx + \int f_\theta(x) \ln \frac{f_{\tilde{\theta}}(x)}{f_{\hat{\theta}(r)}(x)} dx \right]$$

$$= \bar{d}^{(N)}(\theta : \tilde{\theta}) + (1/N)E_r \left[E_x \left[\ln \frac{f_{\tilde{\theta}}(x)}{f_{\hat{\theta}(r)}(x)} \right] \right].$$

The inner expectation in the right-hand side is a function of r and can be shown to be asymptotically distributed as half a χ^2 random variable with Q degrees of freedom, by a generalization of Pearson's χ^2 statistic for the asymptotic distribution of ML estimators [35 §13.4]. The expectation over r of that random variable is $Q/2$, and (53) follows. $\square$

LEMMA 2. *$\tilde{\theta}$ achieves the infimum of $d(\theta : \theta^*)$ over all $\theta^* \in S_Q$.*

Proof. See [21]. $\square$

By application of Lemma 2, we simply denote $d(\theta : \tilde{\theta})$ by $d(\theta : S_Q)$. This quantity represents the directed distance from θ to the sieve set and is independent of the sample size N. For a proper sieve design, $d(\theta : S_Q)$ tends to zero when the sieve grows, that is, Q tends to infinity. As indicated by (C2) in Section 4.3, this is accomplished with the spline sieve.

Lemma 1 has an interesting interpretation. It appears that the estimation error can be decomposed in two terms. The first one is the directed distance from θ to the asymptotic ML estimate in the sieve, also equal to the directed distance from θ to the sieve subset. The second term measures the estimation error within the sieve and is a penalty for estimating Q parameters from a sample of finite size N. This term assumes a particularly simple form, being equal to half the ratio Q/N. Equation (53) shows the very simple tradeoff between the two components of the error. Recognizing this decomposition was a strong motivation for adopting the information distance (52) as a measure of the error. A similar decomposition has been previously recognized by Akaike [1,2] in the derivation of his criterion. However, Akaike's selection of the model order is made from the data themselves, whereas in the sieve design problem this selection depends on the data only via the sample size.

For the radar imaging problem, we let the function $\{S(f, \tau), (f, \tau) \in \Omega\}$ be the parameter θ in the previous discussion. In the following lemma, the proof of which is given in [21], we give an expression for the information distance from S to another function S' in the parameter set. This expression holds under some conditions on the transmitted signal and holds for two important particular cases. In the first case, $N_0 > 0$ and the ambiguity matrix (27) is proportional to the identity matrix. In the second, $N_0 = 0$ and S and s_T are such that the denominator of (54) is bounded away from zero. In the following, we denote by (C) the condition under which Lemma 3 holds.

LEMMA 3. *Assume that either $N_0 > 0$ and for all N, the ambiguity matrix (27) equals $\epsilon_T I_N$, or that $N_0 = 0$, S and S' are bounded away from 0, and all eigenvalues of the matrices $\left(\int_\Omega e^{-j2\pi f(n-m)\Delta t} s_T(n\Delta t - \tau) s_T^*(m\Delta t - \tau) df \, d\tau \right)_{0 \le n, m < N}$ and $\Gamma^\dagger \Gamma$ are positive and bounded away from 0, uniformly in N. Then*

$$(54) \quad \bar{d}(S : S') = |\Omega|^{-1} \int_\Omega \left[-\ln \frac{S(f, \tau) + N_0/\epsilon_T}{S'(f, \tau) + N_0/\epsilon_T} - 1 + \frac{S(f, \tau) + N_0/\epsilon_T}{S'(f, \tau) + N_0/\epsilon_T} \right] df \, d\tau.$$

The information distance (54) attains its minimum (=0) if and only if $S'(f, \tau) = S(f, \tau)$ almost everywhere.

For a one-dimensional version of the estimation problem, in the absence of noise, (54) is recognized as the *Itakura-Saito distance* [15]. This directed distance has been proposed to measure the "distance" of two spectral densities. From a more general standpoint, it has been found that the Itakura-Saito distance is also suitable for measuring the "distance" of two arbitrary positive functions. Suitability means here that this directed distance satisfies certain natural axioms enunciated by Csiszar [10]. Thus, (54) can be viewed as a *generalized Itakura-Saito distance* for measuring the distance between higher-dimensional spectral densities in the presence of noise. Recognizing the essential role of this directed distance is important in the radar imaging problem, and possibly in other areas.

5.2. Mesh size selection.

DEFINITION. A sequence $Q(N)$ leads to estimates $\hat{\theta}_{Q,N}$ which are consistent in information if $\lim_{N \to \infty} d(\theta : \hat{\theta}_{Q,N}) = 0$.

DEFINITION. A sequence $Q(N)$ is optimal in the information sense if, as $N \to \infty$, $d(\theta : \hat{\theta}_{Q,N}) \leq d(\theta : \hat{\theta}_{Q',N})$ for all sequences $Q'(N)$.

The optimal convergence rate of the estimator is obtained by *minimizing the estimation error (53) with respect to Q.* We adopt this minimization procedure as our basic principle for selection of the rate of growth of the sieve. Currently available approaches to mesh size selection are either empirical, or based on other criteria, such as mean-squared error, in the instances where the minimization problem is tractable. Our approach presents the advantage of being more fundamental and less reliant on arbitrary choices. Furthermore, insightful analytical results can be obtained by application of Lemmas 1 and 2.

We evaluate the error measure (53) for the spline sieve (48) which offers a uniform resolution in range and cross-range coordinates. By application of (53),

$$(55) \qquad d^{(N)}(S : \hat{S}) \sim \bar{d}(S : \tilde{S}) + \frac{Q}{2N} \qquad \text{as } N/Q \to \infty,$$

where $Q = (M_{CR} + L - 1)(M_R + L - 1) \sim M_{CR} M_R$ as $M_{CR}, M_R \to \infty$. The convergence rate (55) is to be determined for given range and cross-range resolutions $M_R(N)$ and $M_{CR}(N)$. Following this, the optimal rate of growth of the sieve can be derived by optimizing the convergence rate over all resolutions $M_R(N)$ and $M_{CR}(N)$.

The optimal convergence rate depends on the smoothness properties of the true scattering function. Intuitively we expect fast convergence if the scattering function is smooth and "smooth" splines are used to represent it. Saturation results apply, that is, increasing the smoothness of the spline beyond the degree of smoothness of the function does not improve convergence.

The smoothness of B-splines is determined by their order L. On the other hand, the smoothness of S can be characterized in terms of the space to which the function belongs. For instance, consider the Sobolev space $\mathbf{L}_q^p$. In one dimension, this class is a subspace of the space $\mathbf{L}^p$ where the functions and their derivatives up to order $q - 1$ are absolutely continuous, and the derivatives of order q belong to $\mathbf{L}^p$. In other words, these functions possess q smooth derivatives. The Sobolev spaces play an important role in the study of the approximation power of splines [27 §6]. They also appear in the following results.

Proposition 3 below gives an upper bound on the convergence rate for $\hat{S}$, as M_{CR} and M_R tend to infinity. This upper bound is obtained by performing an analysis of the spline approximation error (54), when the knots are uniformly spaced. The bound is tight in the sense that the constants involved in the final expression can be improved upon, but not the rate of convergence.

PROPOSITION 3. *If Condition (C) is satisfied and if $S \in \mathbf{L}_q^2(\Omega)$, then, for splines of order $L \leq q$, the convergence rate of $\hat{S}$ is upper-bounded by*

$$(56) \quad \hat{d}(S : \hat{S}) \sim \frac{1}{2}\left(\frac{R_{CR}^{(L)}}{M_{CR}^{2L}} + \frac{R_R^{(L)}}{M_R^{2L}} + \frac{M_{CR}M_R}{N} \right) \quad \text{as } M_{CR}, M_R, \frac{N}{M_{CR}M_R} \to \infty,$$

and the estimator is consistent in information if and only if $M_{CR}M_R = o(N)$. The quantities

$$(57a) \qquad R_{CR}^{(L)} := C_L |\Omega|^{-1} \int\limits_{\Omega} \left| \frac{(2f_{\max})^L \partial^L S/\partial f^L}{S(f,\tau) + N_0/\epsilon_T} \right|^2 df\, d\tau$$

and

$$(57b) \qquad R_R^{(L)} := C_L |\Omega|^{-1} \int\limits_{\Omega} \left| \frac{(\tau_{\max})^L \partial^L S/\partial \tau^L}{S(f,\tau) + N_0/\epsilon_T} \right|^2 df\, d\tau$$

that appear in (56) are independent of N, M_{CR} and M_R. We call them the Lth roughness indices of S in cross-range and range, respectively. For splines of order $L > q$, there is saturation of the performance, i.e. $\hat{d}(S : \hat{S})$ is given by (56) in which $L = q$.

Proof. see [21]. ☐

We have given an upper bound for the convergence rate for $\hat{S}$. This expression can now be minimized to determine an approximation to the optimal rate of growth of the sieve.

PROPOSITION 4. *Under the assumptions of Proposition 3, the optimal convergence rate of $\hat{S}$, for a spline of order $L \leq q$, is upper-bounded by*

$$(58) \qquad \hat{d}(S : \hat{S}) = AN^{-L/(L+1)},$$

where $A = \frac{1}{2}\bar{R}^{1/(L+1)}[2(2L)^{-L/(L+1)} + (2L)^{1/(L+1)}]$ is an upper bound on the optimal coefficient. The optimal rate of growth of the sieve is approximated by minimizing (56) over M_{CR} and M_R:

$$(59) \qquad M_{CR,opt} = B_{CR}N^{1/(2L+2)} \quad \text{and} \quad M_{R,opt} = B_R N^{1/(2L+2)},$$

with $B_{CR} = (2L)^{1/(2L+2)}(R_{CR}^{(L)}\bar{R}^{-1/(L+1)})^{1/2L}$, $B_R = (2L)^{1/(2L+2)}(R_R^{(L)}\bar{R}^{-1/(L+1)})^{1/2L}$, and $\bar{R} = (R_{CR}^{(L)}R_R^{(L)})^{1/2}$. *Increasing spline orders L lead to better convergence rates, with saturation for $L \geq q$.*

Proof. The optimal rate of growth of the sieve is obtained by minimizing $d(S : \hat{S})$ over M_{CR} and M_R, for fixed N. Letting M_{CR} and M_R be equal to their optimal values yields (58). The convergence rate, $O(N^{-L/(L+1)}), L \leq q$, improves with increasing L. It follows that the optimal spline order is q. □

The results presented in Propositions 3 and 4 indicate that the estimates obtained with our sieve are consistent in information. The optimal convergence rate is obtained when the order of the spline is equal to the "degree of smoothness" of the scattering function. Smoother functions lead to better convergence rates.

The approach described in this section makes use of a priori information on the smoothness properties of the scattering function. Even if such information is not available, consistent estimates can be obtained by letting $M_{CR}M_R = o(N)$. Naturally, the convergence rates achieved in this fashion are not as good.

6. Discrete processing.

In this section, we propose a numerically efficient version of the EM algorithm derived in Section 4.2.4. We recommend an estimation based on the discrete model as an approximation to the true model, with $I_{CR}I_R = N$ samples in Ω only. This particular discretization leads to a tremendous simplification of the estimation algorithm of the previous section. The resulting estimator is consistent in information and even offers the same convergence rate as the estimator for the true model.

6.1. Convergence rate for discrete model. Consider the discrete model (19) as an approximation to the true model, with $I_{CR}I_R = N$. Discrete estimates $\{\hat{\sigma}^2(k,i), 0 \leq k < I_{CR}, 0 \leq i < I_R\}$ for the scattering function can be obtained using the sieve (34). For Λ_Q fixed, let $\hat{S}$ be the ML estimate in the sieve for the continuous model (32). Under condition (C) of Section 5.1, it can be shown that [21]

$$(60) \qquad d^{(N)}(S : \hat{\sigma}) = d^{(N)}(S : \hat{S}) + O(N^{-1}) \text{ as } N \to \infty.$$

Thus, the discrete approximation introduces an error $O(N^{-1})$ under the information distance. This error is dominated by the first term in the right-hand side, which was shown in (58) to be $O(N^{-L/(L+1)})$. We conclude that the information performance of the discrete estimator is identical to that of the continuous estimator, and in particular, that

(1) The discrete estimator is consistent in information when $M_{CR}M_R = o(N)$.

(2) Optimal convergence rates of the discrete estimator are obtained for M_{CR} and $M_R = O(N^{1/(2L+2)})$, with saturation for $L \geq q$. Then, M_{CR} and M_R are $O(N^{1/(2L+2)})$, and the convergence rate is $O(N^{-L/(L+1)})$.

6.2. Estimation algorithm. The covariance matrices K_r and K_m for the discrete model are given in (21) and (38), respectively. When condition (C) of

Section 5.1 holds, K_r, like K_m, can be written as the product of three matrices, the middle one being diagonal:

$$(61) \qquad\qquad K_r = \Gamma^\dagger(\Sigma + N_0/\epsilon_T I_N)\Gamma.$$

Then, the first update equation (45) becomes

$$(62) \qquad\qquad \Sigma^{(p)} = \sum_{j \in \Lambda_Q} \hat{a}(j)^{(p)} \Psi_j.$$

Next, introducing the expressions (38), (46) and (61) in the update equation (47), we obtain for the trace in the right-hand side

(63)
$$Tr\left[K_m K_r^{(p)-1}(rr^\dagger - K_r^{(p)})K_r^{(p)-1}\right]$$

$$= Tr\left[\Gamma^\dagger \Psi_m(\Sigma^{(p)} + N_0/\epsilon_T I_N)^{-1}((\Gamma^\dagger)^{-1}rr^\dagger\Gamma^{-1} - \Sigma^{(p)} - N_0/\epsilon_T I_N)\right.$$

$$\left. \times (\Sigma^{(p)} + N_0/\epsilon_T I_N)^{-1}(\Gamma^\dagger)^{-1}\right]$$

$$= Tr\left[\Psi_m(\Sigma^{(p)} + N_0/\epsilon_T I_N)^{-1}(pp^\dagger - \Sigma^{(p)} - N_0/\epsilon_T I_N)(\Sigma^{(p)} + N_0/\epsilon_T I_N)^{-1}\right]$$

$$= \sum_{k,i \in D_m} \frac{\psi_m^{(I_{CR},I_R)}(k,i)}{(\sigma^2(k,i)^{(p)} + N_0/\epsilon_T)^2}\left(|p(k,i)|^2 - \sigma^2(k,i)^{(p)} - N_0/\epsilon_T\right),$$

where we have introduced the preimage (29) obtained from a linear preprocessing of the data. Another update equation is obtained from (47) and (63):

(64)
$$\hat{a}(m)^{(p+1)} = \hat{a}(m)^{(p)} + \frac{1}{|\mathbf{D}_m|}\left(\hat{a}(m)^{(p)}\right)^2 \sum_{k,i \in \mathbf{D}_m} \frac{\psi_m^{(I_{CR},I_R)}(k,i)}{(\sigma^2(k,i)^{(p)} + N_0/\epsilon_T)^2}$$

$$\times \left(|p(k,i)|^2 - \sigma^2(k,i)^{(p)} - N_0/\epsilon_T\right), \quad m \in \Lambda_0.$$

Equations (62) and (64) are the update equations at stage p of the EM algorithm for the discrete model with $I_{CR}I_R = N$, when condition (C) holds. These equations take a remarkably simple form. In (47), the matrix operations that have to be performed to evaluate the trace, namely inversion and multiplications, are computationally costly. In (64), these operations are performed on diagonal matrices. This leads to a tremendous reduction in complexity. Notice that the covariance for p is $E[pp^\dagger] = \Sigma + N_0(\Gamma\Gamma^\dagger)^{-1}$. Thus, under the assumptions made in this section, (29) defines an orthogonalization of the observations. This orthogonalization leads to a substantial reduction in complexity of the ML estimation problem. As shown by (60), this is achieved at no cost on the convergence rate of the estimator.

6.3. Algorithm complexity. In general, the complexity of the preprocessing (29) is N^2. However, for certain types of transmitted signals such as stepped-frequency waveforms, Γ is a two-dimensional DFT matrix; in this case, the complexity of the preprocessing is only $N \log_2 N$.

We now discuss the computational complexity for the update equations (62) and (64) at each step of the EM algorithm. The number of nonzero samples $\psi_m^{(I_{CR}, I_R)}(k, i)$ in (62) and (64) is equal

to $|\mathbf{D}_m|$. Thus, (62) involves $\sum_{m \in \Lambda_Q} |\mathbf{D}_m|$ additions and (64) the same number of multiplications and additions. For the spline sieve (48), the overlapping factor ρ of the elementary functions is equal to L^2, so that $\sum_{m \in \Lambda_Q} |\mathbf{D}_m| = L^2 I_{CR} I_R = L^2 N$. In this instance, the complexity of the update equations is simply proportional to N and the proportionality constant is the overlapping factor of the splines. This factor is usually relatively small. For other types of elementary functions, the conclusions can be quite different. For instance, when the elementary functions have global support, as occurs with Fourier representations, ρ is equal to the number Q of elementary functions, which can be very large. From a numerical standpoint, the benefits of a system of locally-supported elementary functions are clear.

The complexity of the operations at each step of the EM algorithm should be compared with that obtained for the exact model in Section 4.2.4, that is, $N^3 + QN^2$. Let us give a numerical example. For a 128×128 pixel image represented via biquadratic splines, $N \sim 1.6 \times 10^4$ and $L = 3$. Then, $\rho = 9$ and the complexity of the update equations is only 1.4×10^5 instead of 4×10^{12}.

From the discussion above, the global complexity of the estimation algorithm is $N^2 + N_{EM} \rho N$ or $N \log_2 N + N_{EM} \rho N$, where N_{EM} is the number of iterations of the EM algorithm required for convergence of the estimates and ρ is the overlapping factor of the elementary functions. The first term in both expressions is the contribution of the preprocessing of the data and the second is the complexity of the EM algorithm.

7. Simulation results.

In order to visualize the performance of the estimation algorithm, we have performed a series of simulations for a phantom target. This target is a sphere rotating about a point along the line of sight. For this simple object, the scattering function can be calculated [33]. A discrete version ($I_{CR} I_R = 128 \times 128$ pixels) of this image is displayed in Figure 1a. Independent samples of the discrete reflectivity process are produced using a Gaussian random number generator. Each of these samples is zero-mean and has variance equal to the corresponding sample of the scattering function image. The transmitted signal is a stepped-frequency waveform [34] for which the parameters are designed so that the ambiguity matrix (27) is the identity. The return echo is generated from the radar equation (19), with $N = I_{CR} I_R = 128^2$.

In Figures 1b-e, we present different estimates of the scattering function, in the presence of additive noise with variance equal to 60. The maximal intensity of the scattering function is equal to 300 and its mean value to 7. Figure 1b shows the conventional estimate $|p|^2$ from (29) discussed in Section 3. Each pixel of this image is exponentially distributed. The rough aspect of the image is known as *speckle noise*.

Figures 1c-e show three maximum likelihood sieve-constrained estimates. The image region Ω is partitioned uniformly. The images show spline orders of 1, 2, and 3, respectively. Since the scattering function for the rotating sphere is smooth, better estimation performance is obtained using higher-order splines, as can be seen from Figures 1c-e.

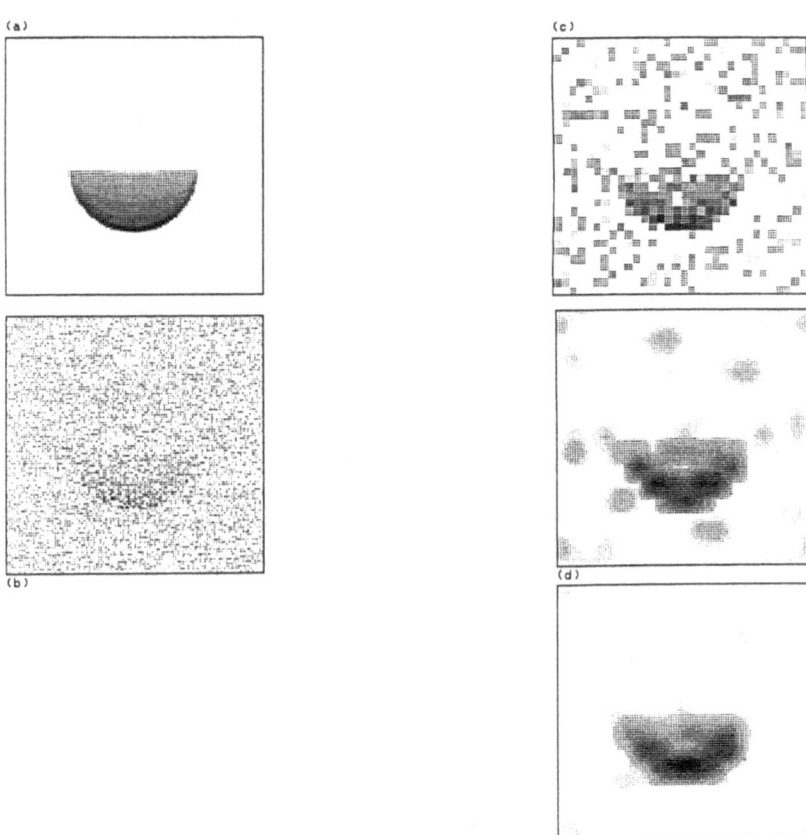

Figure 1. Estimates of the Scattering Function ($N_0 = 60$)

(a) Scattering function for the rotating sphere.
(b) Conventional estimate.
(c) Maximum-likelihood estimate, $L = 1, M_{CR} = M_R = 32$.
(d) Maximum-likelihood estimate, $L = 2, M_{CR} = M_R = 16$.
(e) Maximum-likelihood estimate, $L = 3, M_{CR} = M_R = 16$.

This observation is consistent with the results in Propositions 3 and 4. For all three splines, the selection of the resolution level was done on the basis of a subjective, visual criterion. Another approach would be estimating the roughness indices (57) for the scattering function and using these estimates to compute the resolution level in both coordinates from (59). We have not attempted to follow such a strategy, so the quality of the estimates that would be produced in this fashion is unknown. Clearly, the ML estimates are better than the conventional estimates.

Other experiments have been performed which emphasize the multiresolution nature of the algorithm. These are reported in [22]. It is our intent to use these methods on actual radar data and report those results in the future.

8. Conclusions.

A method of sieves based on polynomial splines has been proposed for radar imaging. Statistical assumptions on radar reflections and additive noise are made which transform the problem of radar imaging to a statistical estimation problem. This problem is ill-posed, so sieve constraints are introduced to regularize the estimates. As the size of the data set grows, the sieve set is allowed to grow. Asymptotic properties of the resulting estimates are derived, including consistency in information.

A generalization of the method used here is introduced in [22]. In that paper, the multiresolution character of the algorithm is emphasized and applications to estimating functions of an arbitrary number of variables are introduced.

There are some limitations to the approach. As presented here, the method relies on the assumption of Gaussian reflections. Other models for reflections are possible [31], and other algorithms would apply in those cases. The approach we would take under other models would be to derive the likelihood function for the available data and to maximize it over a sieve set. The model used here is often referred to as a speckle model. Currently, we are exploring appropriate models for glint reflections and their use in estimation problems.

The choice of polynomial splines as our sieve constraints was motivated by several factors. First, the enforcement of the nonnegativity of the scattering function reduces to requiring the parameters in the sieve set to be positive. Enforcing nonnegativity for basis functions which take on negative values can be complicated. Second, sieves naturally allow estimation at various resolution levels. This may be accomplished on uniform partitions by just changing the size of the partition, or it may be accomplished nonuniformly by selecting different partition sizes in different parts of the image. How to perform this second approach in a theoretically meaningful manner is a difficult problem. Other reasons for choosing splines are outlined in Section 4.

REFERENCES

[1] H. AKAIKE, *A New Look at the Statistical Model Identification*, IEEE Transactions on Automatic Control, Vol. 19, No. 6 (1974), pp. 716-723.

[2] H. AKAIKE, *On Entropy Maximization Principle*, in: *Applications of Statistics*, (P.R. Krishnaiah, ed.), North-Holland (1977), pp. 27-41.

[3] T. W. ANDERSON, *Estimation of Covariance Matrices Which Are Linear Combinations or Whose Inverses Are Linear Combinations of Given Matrices*, Chapter 1, in: *Essays in Probability and Statistics*, (R.C. Bose et al., Eds.), Chapel Hill, NC: University of North Carolina Press (1970), pp. 1-24.

[4] A. R. BARRON AND C. H. SHEU, *Approximation of Density Functions by Sequences of Exponential Families*, University of Illinois at Urbana-Champaign, Department of Statistics Technical Report # 8, submitted to Annals of Statistics (1989).

[5] G. L. BRETTHORST, *Bayesian Spectrum Analysis and Parameter Estimation*, Lecture Notes in Statistics, Springer Verlag, New York (1988).

[6] J. P. BURG, *Maximum-Entropy Spectral Analysis*, Proceedings 37th Meeting Society of Exploration Geophysicists Oklahoma City, OK (October 1967).

[7] J. P. BURG, D. G. LUENBERGER, AND D. L. WENGER, *Estimation of Structured Covariance Matrices*, Proceedings IEEE, Vol. 70, No. 9 (1982), pp. 963-974.

[8] Y. CHOW AND U. GRENANDER, *A Sieve Method for the Spectral Density*, Annals of Statistics, Vol. 13, No. 3 (1985), pp. 998-1010.

[9] R. COGBURN AND H. T. DAVIS, *Periodic Splines and Spectral Estimation*, Annals of Statistics, Vol. 2, No. 6 (1974), pp. 1108-1126.

[10] I. CSISZAR, *Why Least-Squares and Maximum-Entropy?*, Mathematical Institute of the Hungarian Academy of Science Preprint No. 19/1989, submitted to IEEE Transactions on Information Theory (1990).

[11] A. D. DEMPSTER, N. M. LAIRD, AND D. B. RUBIN, *Maximum-Likelihood from Incomplete Data via the EM Algorithm*, Journal of the Royal Statistical Society, Vol. B39, No. 1 (1977), pp. 1-37.

[12] J. L. DOOB, *Stochastic Processes*, Wiley, New York (1953).

[13] U. GRENANDER, *Abstract Inference*, Wiley, New York (1981).

[14] P. J. HUBER, *The Behavior of Maximum Likelihood Estimates under Nonstandard Conditions*, Proceedings 5th Berkeley Symposium on Mathematical Statistics and Probability University of California Press, Vol. 1 (1967), pp. 221-233.

[15] F. ITAKURA AND S. SAITO, *Analysis-Synthesis Telephony Based on the Maximum Likelihood Method*, Proceedings 6th Int. Conf. Acoust. C, Tokyo, Japan (1968), pp. 17-20.

[16] L. K. JONES, *Approximation-Theoretic Derivation of Logarithmic Entropy Principles for Inverse Problems and Unique Extension of the Maximum Entropy Method to Incorporate Prior Knowledge*, SIAM Journal on Applied Mathematics, Vol. 49, No. 2 (1989), pp. 650-661.

[17] S. M. KAY AND S. L. MARPLE, *Spectrum Estimation - A Modern Perspective*, Proceedings IEEE, Vol. 69, No. 11 (1981), pp. 1380-1419.

[18] D. L. LUENBERGER, *Introduction to Linear and Nonlinear Programming*, Addison Wesley, Reading, MA (1973).

[19] M. I. MILLER, K. B. LARSON, J. E. SAFFITZ, D. L. SNYDER, AND L. J. THOMAS, JR., *Maximum Likelihood Estimation Applied to Electron-Microscopic Autoradiography*, Journal of Electron Microscopy Technique, Vol. 2 (1985), pp. 611-636.

[20] M. I. MILLER AND D. L. SNYDER, *The Role of Likelihood and Entropy in Incomplete-Data Problems: Applications to Estimating Point-Process Intensities and Toeplitz Constrained Covariances*, Proceedings IEEE, Vol. 75, No.7 (1987), pp. 892-907.

[21] P. MOULIN, *A Method of Sieves for Radar Imaging and Spectrum Estimation*, D.Sc. Dissertation, Washington University, Saint Louis, Missouri (1990).

[22] P. MOULIN, J. A. O'SULLIVAN, AND D. L. SNYDER, *A Method of Sieves for Multiresolution Spectrum Estimation and Radar Imaging*, to appear in IEEE Transactions on Information Theory (1992).

[23] H. NAPARST, *Radar Signal Choice and Processing for a Dense Target Environment*, in *Signal Processing Part II: Control Theory and Applications*, (F. A. Grunbaum, J. W. Helton, and P. Khargonecker eds.), Springer-Verlag, New York (1990), pp. 293-319.

[24] J. A. O'SULLIVAN AND D. L. SNYDER, *High Resolution Radar Imaging Using Spectrum Estimation Methods*, in Signal Processing Part II: Control Theory and Applications, (F. A.

Grunbaum, J. W. Helton, and P. Khargonecker eds.), Springer-Verlag, New York (1990), pp. 335–346.

[25]   H. L. ROYDEN, *Real Analysis*, McMillan, New York (1968).

[26]   I. J. SCHOENBERG, *Contributions to the Problem of Approximation of Equidistant Data by Analytic Functions, Part B: On the Problem of Osculatory Interpolation, a Second Class of Analytic Approximation Formulae*, Quarterly of Applied Mathematics, Vol. 4 (1946), pp. 112–141.

[27]   L. L. SCHUMAKER, *Splines Functions: Basic Theory*, Wiley, New York (1981).

[28]   D. L. SNYDER AND M. I. MILLER, *The Use of Sieves to Stabilize Images Produced with the EM Algorithm for Emission Tomography*, IEEE Transactions on Nuclear Science, Vol. 32 (1985), pp. 3864–3872.

[29]   D. L. SNYDER, M. I. MILLER, L. J. THOMAS, AND D. G. POLITTE, *Noise and Edge Artifacts in Maximum-Likelihood Reconstructions for Emission Tomography*, IEEE Transactions on Medical Imaging, Vol. 28, No. 3 (1987), pp. 228–238.

[30]   D. L. SNYDER, J. A. O'SULLIVAN, AND M. I. MILLER, *The Use of Maximum-Likelihood Estimation for Forming Images of Diffuse Radar-Targets from Delay-Doppler Data*, IEEE Transactions on Information Theory, Vol. 35, No. 3 (1989), pp. 536–548.

[31]   P. SWERLING, *Probability of Detection for Fluctuating Targets*, IRE Transactions on Information Theory, Vol. IT-6 (1960), pp. 269–308.

[32]   A. TIKHONOV AND V. ARSENIN, *Solutions of Ill-Posed Problems*, Winston and Sons, Washington, D.C. (1977).

[33]   H. L. VAN TREES, *Detection, Estimation, and Modulation Theory, Part III*, Wiley and Sons, New York (1971).

[34]   D. R. WEHNER, *High Resolution Radar*, Artech House, Norwood, MA (1987).

[35]   S. S. WILKS, *Mathematical Statistics*, Wiley, New York (1962).

SOME PROBLEMS IN OBTAINING AND USING INCREMENTAL DIFFRACTION COEFFICIENTS

ROBERT A. SHORE* AND ARTHUR D. YAGHJIAN*

Abstract. This paper describes a recent and highly useful method for obtaining incremental diffraction coefficients for an important class of canonical scatterers. If a closed form expression can be supplied for the scattered far field of a two-dimensional planar scatterer, the incremental diffraction coefficients at arbitrary angles of incidence and scattering can be found immediately through direct substitution into general expressions. No integration, differentiation, or specific knowledge of currents is required. The direct substitution method for determining incremental diffraction coefficients is, however, limited to perfectly conducting scatterers that consist of planar surfaces, such as the wedge, the slit in an infinite plane, the strip, parallel or skewed planes, polygonal cylinders, or any combination thereof; and requires a closed-form expression (exact or approximate) for the two-dimensional scattered far field produced by the current on each different plane of the canonical scatterer. The usefulness of incremental diffraction coefficients is demonstrated by presenting a few applications to scattering problems, showing the importance of the nonuniform current contribution to the scattered fields. Examples are shown for the calculation of reflector antenna patterns, and for the calculation of the field scattered by a perfectly conducting circular disk illuminated by a plane wave.

I. Introduction.

Incremental diffraction coefficients provide an important technique for enhancing the accuracy of the physical optics (PO) approximation. The PO approximation, widely used for calculating scattering from perfectly conducting bodies, consists of approximating the actual currents at a surface point by those induced on an infinite perfectly conducting plane tangent to the body at the point. The PO approximation works well away from shadow boundaries provided that the radii of curvature of the reflecting surface are large compared with the wavelength, and that there are no surface discontinuities. When shadow boundaries or surface discontinuities are present, the accuracy of the scattered fields obtained via the PO approximation can be significantly improved if the fields radiated by the nonuniform currents (the difference between the actual and PO currents) can be closely approximated and included in the scattering calculations. The far-field contribution from the nonuniform current of a differential element of an edge or shadow boundary – the incremental diffraction coefficient (IDC) – can be obtained by regarding the edge or shadow boundary locally as the edge or shadow boundary of a canonical scatterer (e.g., wedge, half-plane, cylinder) provided that an expression is available for the IDC of the canonical scatterer. The IDC can then be integrated along the edges or shadow boundaries to obtain the far field of the nonuniform currents. When IDC's can be found they provide a powerful technique for augmenting the accuracy of the fields calculated using the PO approximation. In contrast to other techniques commonly employed to account for scattering from edge discontinuities of perfectly conducting surfaces, the integration of IDC's along edge discontinuities yields corrections to PO fields that are valid in virtually all ranges of pattern angles, and so avoids the need for several distinct formulations, each valid in a particular range.

*Rome Laboratory, Hanscom AFB, MA 01731.

In this paper after a brief introduction to IDC's, we describe a recent and highly useful method for obtaining IDC's for an important class of canonical scatterers. If a closed form expression can be supplied for the scattered far field of a two-dimensional planar scatterer, the IDC's at arbitrary angles of incidence and scattering can be found immediately through direct substitution in general expressions. No integration, differentiation, or specific knowledge of currents is required. The direct substitution method for determining IDC's is, however, limited to perfectly conducting scatterers that consist of planar surfaces, such as the wedge, the slit in an infinite plane, the strip, parallel or skewed planes, polygonal cylinders, or any combination thereof; and requires a closed-form expression (whether exact or approximate) for the two-dimensional scattered far field produced by the current on each different plane of the canonical scatterer.

We then demonstrate the utility of IDC's by presenting a few applications to scattering problems, showing the importance of the nonuniform current contribution to the scattered fields. Examples are shown for the calculation of reflector antenna patterns, and for the calculation of the field scattered by a perfectly conducting circular disk illuminated by a plane wave.

Finally, as an invitation to further work in this area, we draw attention to the importance of removing the restriction of the substitution method for obtaining IDC's to planar, perfectly conducting scatterers. It would be of much utility if a simple method not involving current integration could be found for obtaining IDC's for a circular cylinder or a parabolic cylinder (rounded wedge), or for nonperfectly conducting scatterers. In addition, the attractiveness of using IDC's for calculating scattering from perfectly conducting scatterers with discontinuities that can be locally modeled by planar canonical scatters would be considerably enhanced if a general purpose computer program could be developed for such scatterers that would circumvent the currently rather lengthy procedure of transforming from the local coordinate system of the IDC's to the global coordinate system of the scatterer in order to integrate the IDC's.

II. Basics of Incremental Diffraction Coefficients.

Incremental diffraction coefficients arise in an attempt to improve upon the accuracy of the physical optics method of calculating scattering from perfectly conducting objects. In the PO method the current at a point of the illuminated region of the scatterer is approximated by the current that would be induced by the incident field on an infinite perfectly conducting plane tangent to the scattering surface at the point. That is,

$$(1) \qquad \overline{K}_{PO} = 2\hat{n} \times \overline{H}_i$$

where $\overline{H}_i$ is the incident magnetic field and $\hat{n}$ is the unit normal to the surface at the point of incidence. The current over the shadow region of the surface is assumed to be zero. The PO approximation of the surface current works well away from shadow boundaries provided that the radii of curvature of the reflecting surface are large compared with the wavelength, and that there are no surface discontinuities. In regions of the scatterer where there are surface discontinuities or shadow boundaries, it is to be expected that the PO current approximation will not be satisfactory.

The difference between the actual total current and the current as given by the PO approximation is called the nonuniform current. The nonuniform current can have a considerable effect on the scattered fields, being especially important for the accurate determination of cross-polarized fields, sidelobe fields, and fields near nulls. The nonuniform current is strongest near the regions of discontinuities and shadow boundaries and usually decays rapidly away from them. If we knew what the nonuniform currents were, we could then integrate them, multiplied by the free-space Green's function, to obtain a far field which could then be added to the PO far field to obtain the total far field. This is a big "if", however, because the nonuniform currents are very complicated and difficult to obtain. But for purposes of understanding what IDC's are, let us assume for now that the nonuniform currents can be found. The integration over the scattering surface is a two-dimensional integration. For the example of a perfectly conducting half-plane illuminated by a plane wave, let us take one of the directions of integration – the z-direction – to be along the edge of the half-plane, and the other direction – the x-direction – to make a given angle with the edge (see Figure 1). The vector potential, $\overline{A}$, corresponding to the nonuniform current, $\overline{\mathcal{K}}_{nu}(x, z)$, is given by

$$(2) \qquad \overline{A} = \frac{1}{4\pi} \int_{-\infty}^{\infty} dz' \int_0^{\infty} dx' \overline{\mathcal{K}}_{nu}(x', z') \frac{e^{ik|\overline{r}-\overline{r}'|}}{|\overline{r} - \overline{r}'|}$$

where $\overline{r}$ and $\overline{r}'$ are the far-field point and integration point respectively. The far-field scattered magnetic field due to the nonuniform current is then

$$(3) \qquad \overline{H}_s(\overline{r}) = \nabla \times \overline{A}.$$

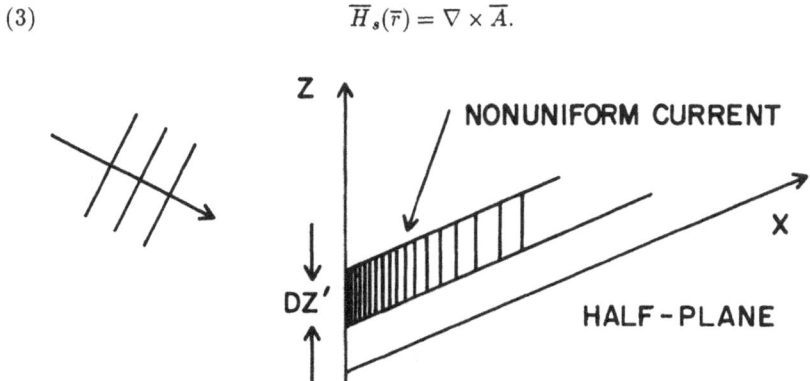

Figure 1. Differential nonuniform current strip at edge of half-plane.

If we perform the surface integration in (2) along the transverse (x) direction first, then we obtain for each differential element, dz', of the edge of the half-plane, a differential far field

$$(4) \qquad d\overline{H}_s(\overline{r}, z') = \frac{dz'}{4\pi} \nabla \times \int_0^\infty dx' \overline{\mathcal{K}}_{nu}(x', z') \frac{e^{ik|\overline{r}-\overline{r}'|}}{|\overline{r}-\overline{r}'|}.$$

This differential far field is the nonuniform current IDC. By integrating the IDC along the edge of the half-plane, we obtain the nonuniform far field. Obviously we can generalize this idea to any surface discontinuity or shadow boundary. The IDC will, in general, depend on the shape of the illuminated object, especially near the discontinuity or shadow boundary, on the direction of incidence and the polarization of the illuminating field, and on the direction of the far-field point. It is important to note that the IDC is not unique, because it depends on the choice of the transverse integration direction.

The reason for the interest in, and importance of, determining IDC's for canonical shapes such as the wedge, lies in the fact that the nonuniform current at a given point on some particular electrically large scatterer can often be well approximated by the nonuniform current on a corresponding canonical scatterer that conforms to the shape of the particular scatterer in the vicinity of the given point. For example, suppose we have a thin reflecting surface of arbitrary shape. The nonuniform currents excited on a differential strip of surface transverse to the edge can be closely approximated near the edge by those excited on a strip of a half-plane tangent to the particular reflecting surface at the edge point (see Figure 2). To obtain an approximation to the far field due to the nonuniform current everywhere on the thin reflecting surface, we simply integrate the half-plane IDC around the edge of the reflecting surface. And so, in general, if we have a computer program for calculating the PO far field of a particular scattering object, and if we know the IDC for a canonical scatterer that conforms to the shape of the particular object at points on a curve of discontinuity, we can then enhance the accuracy of the PO far field by integrating the canonical IDC along the curve of discontinuity of the object.

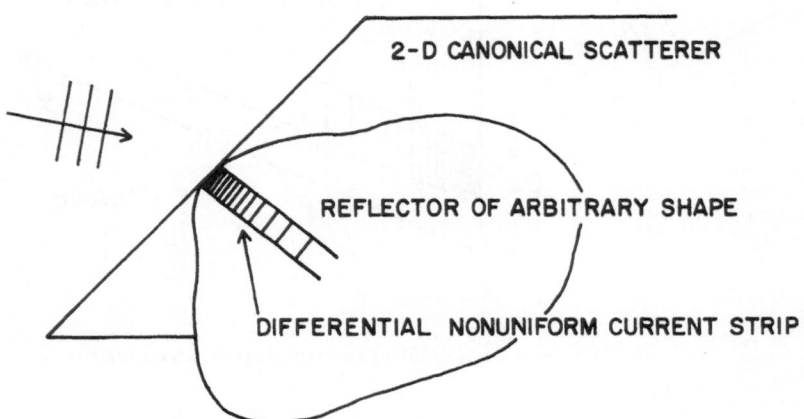

Figure 2. Modelling particular scatterers at discontinuities by canonical scatterers.

Obtaining IDC's by the procedure of finding and integrating the nonuniform current is not an easy matter. In fact, prior to our work they had been obtained for only one type of scatterer: the perfectly conducting wedge.[1-3] Rather surprisingly, however, we found[4] that IDC's can in some cases be determined without any knowledge of currents whatsoever. For an important class of two-dimensional perfectly conducting scatterers consisting of planar surfaces, such as the wedge, the strip, the slit, parallel or skewed planes, and polygonal cylinders, all that is necessary to determine the IDC's is an expression for the far field produced by the current on each different plane. The IDC for the scatterer can then be obtained by a simple substitution of the far field of each of the planes into a general expression and then summing the results. Knowledge of currents is not required nor is integration needed. In the following section we give the derivation of this substitution method for obtaining IDC's because the derivation makes clear why the method is apparently limited to the class of perfectly conducting planar canonical scatterers that we have just mentioned.

III. IDC's in Terms of Cylindrical Far Fields.

Consider a planar facet of a two-dimensional perfectly electrically conducting scatterer in free space (see Figure 3). The scatterer extends uniformly to infinity in the z-direction, and the planar facet lies in the xz-plane. The scatterer is illuminated by a plane wave with propagation constant $\overline{k}$ whose direction is given by the spherical polar angles θ_o and ϕ_0. The magnitude of k equals $2\pi/\lambda = \omega/c$ where λ is the free-space wavelength, ω is the angular frequency of the suppressed $\exp(-i\omega t)$ time dependence, and c is the speed of light in free space.

Maxwell's equations and the boundary condition of zero tangential electric field on the scatterer show that for plane wave illumination the surface current $\overline{\mathcal{K}}(\overline{r}')$ at any point $\overline{r}'$ on the current sheet can be factored as

$$(5) \qquad \overline{\mathcal{K}}(\overline{r}') = \overline{e}^{ikz'\cos\theta_0}\overline{K}(x').$$

That is, the z'-dependence can be written explicitly in terms of the elevation angle of the incident plane wave, and thus the remaining current factor $\overline{K}(x')$ depends on the transverse coordinate x', but not on the longitudinal coordinate. Of course $\overline{K}(x')$ also depends implicitly on the angles θ_0 and ϕ_0 defining the direction of propagation of the incident plane wave, and on the polarization of the incident wave, and is proportional to the complex magnitude of the incident wave. Also $\overline{\mathcal{K}}$ need not refer to the total current; it can equally well refer to the nonuniform current or the PO current.

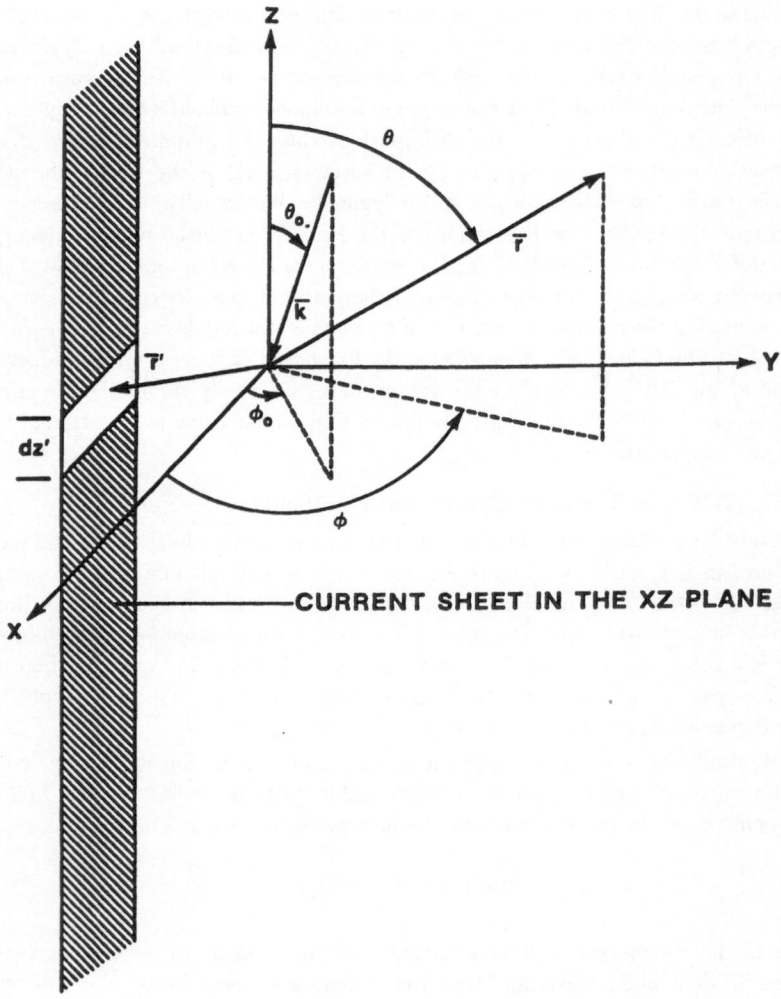

Figure 3. Planar Facet of a Two-Dimensional, Perfectly Conducting Scatterer in Free Space.

It is straightforward to express the IDC $\overline{dH}_s(\overline{r})$, the scattered magnetic far field at the point $\overline{r}$ radiated by the incremental current strip of width dz', in terms of the current by taking the curl of the vector potential and specializing the result to the far field thereby obtaining

$$(6) \qquad \overline{dH}_s(\overline{r}) \overset{r \to \infty}{\sim} dz' \frac{e^{ikr} ik}{4\pi r} \hat{r} \times \overline{A} \equiv dz' C \hat{r} \times \overline{A}$$

with

(7)
$$\overline{A} = \int_{-\infty}^{\infty} \overline{K}(x')e^{-ikz' \sin \theta \cos \phi} dx'$$

and

(8)
$$C = \frac{e^{ikr} ik}{4\pi r}.$$

(The z'-dependence of the incremental far field is eliminated by choosing the incremental current sheet at $z' = 0$.) So (6) is an expression for the IDC we are looking for in terms of the vector potential $\overline{A}$ which, of course, is not yet known because it is given in terms of the unknown current.

Now the cylindrical far field of the entire current sheet can be found by integrating the expression for the differential field (prior to its specialization in (6) to the far field) from $z' = -\infty$ to $+\infty$ and then making the far-field approximation, thereby obtaining

(9)
$$\overline{H}_s(\overline{r}) \stackrel{r \to \infty}{\sim} \frac{e^{i\pi/4} e^{ik\hat{r}_0 \cdot \hat{r}} ik}{(8\pi k\rho \sin \theta_0)^{1/2}} \hat{r}_0 \times \overline{A}_0 \equiv C_0 \hat{r}_0 \times \overline{A}_0$$

where

(10)
$$\overline{A}_0 = \int_{-\infty}^{\infty} \overline{K}(x')e^{-ikz' \sin \theta_0 \cos \phi} dx'$$

and

(11)
$$C_0 = \frac{e^{i\pi/r} e^{ik\hat{r}_0 \cdot \hat{r}} ik}{(8\pi k\rho \sin \theta_0)^{1/2}}.$$

In (9) $\hat{r}_0$ is defined to be the unit vector $\hat{r}$ evaluated at $\theta = \pi - \theta_0$ and ϕ. Thus the magnetic far field is of the form of a cylindrical wave propagating along the generators of the diffraction cone – the scattered waves that make an angle with the longitudinal z-axis equal to $\pi - \theta_0$.

If we compare the integrals, (7) and (10), defining $\overline{A}$ and $\overline{A}_0$, we see that the two integrals differ only in that $\sin \theta$ appears in the integral for $\overline{A}$ and $\sin \theta_0$ in the integral for $\overline{A}_0$. Hence $\overline{A}$ can be obtained from $\overline{A}_0$ by making the substitution

(12)
$$\cos \phi \to \frac{\sin \theta}{\sin \theta_0} \cos \phi$$

or

(13)
$$\phi \to \alpha = \cos^{-1} \left(\frac{\sin \theta}{\sin \theta_0} \cos \phi \right).$$

Thus

(14)
$$\overline{A}(\theta, \phi) = \overline{A}_0 \left(\cos \phi \to \frac{\sin \theta}{\sin \theta_0} \cos \phi \right).$$

If we can find $\overline{A_0}$ we will then also know $\overline{A}$ and hence the IDC $\overline{dH}_s(\overline{r})$. At this point it may be asked why it is not possible to simply substitute $\sin\theta$ for $\sin\theta_0$ in the expression for $\overline{A_0}$ to obtain A if $\overline{A_0}$ is known. The answer is that the current sheet $\overline{K}(x')$ depends implicitly on the angle θ_0, the polar angle of the illuminating plane wave propagation vector, so that if θ_0 is changed so is $\overline{K}(x')$, whereas if ϕ is changed there is no change in $\overline{K}$ because ϕ is the azimuth angle of the field point.

We now assume a known closed-form expression for the far field $\overline{H}_s(\overline{r})$ of the planar face of the canonical scatterer (e.g., a face of a wedge). In the far field $\overline{H}_s(\overline{r})$, in spherical coordinates, has only θ- and ϕ-components. Thus we are given $H_{s\theta}$ and $H_{s\phi}$ such that

$$(15) \qquad \overline{H}_s(r) \overset{r\to\infty}{\sim} H_{s\theta}\hat{\theta}_0^\pi + H_{s\phi}\hat{\phi}$$

where $\hat{\theta}_0^\pi$ is θ evaluated at $\pi - \theta_0$. Because the current sheet by assumption is in the xz-plane, $K_y = 0$ and so $\overline{A_0}$ has x- and z-components only. Equating (15) to (9) expanded in spherical components we obtain

$$(16) \qquad H_{s\theta}\hat{\theta}_0^\pi + H_{s\phi}\hat{\phi} = C_0 \left[A_{0x}\sin\phi\,\hat{\theta}_0^\pi - (A_{0x}\cos\theta_0\cos\phi + A_{0z}\sin\theta_0)\,\hat{\phi} \right]$$

from which we can solve for A_{0x} and A_{0z} in terms of $H_{s\theta}$ and $H_{s\phi}$:

$$(17) \qquad A_{0x} = \frac{1}{C_0} H_{s\theta}\sin\phi$$

and

$$(18) \qquad A_{0z} = -\frac{1}{C_0}\left(H_{s\theta}\cot\theta_0\cot\phi + \frac{H_{s\phi}}{\sin\theta_0} \right).$$

It is this step of solving for the two cartesian components of the vector potential of the two-dimensional far field that is made possible by the fact that we are working with a planar facet of a scatterer so that $\overline{A_0}$ has only two cartesian components that can be solved for from the two equations obtained by equating the coefficients of the two spherical unit vectors. If the surface were curved $\overline{A_0}$ would have three cartesian components and so we would be unable to solve for them.

Having solved for the two cartesian components of $\overline{A_0}$, we have immediately by the change of variables, (12) or (13), found above that relates $\overline{A}$ to $\overline{A_0}$, the two cartesian components of $\overline{A}$:

$$(19) \qquad A_x(\theta, \phi) = \frac{1}{C_0} H_{s\theta}(\phi \to \alpha)\sin\alpha$$

and

$$(20) \qquad A_z(\theta, \phi) = -\frac{1}{C_0 \sin\theta_0}\left[H_{s\theta}(\phi \to \alpha)\frac{\cot\theta_0\sin\theta\cos\phi}{\sin\alpha} + H_{s\phi}(\phi \to \alpha) \right]$$

where

(21)
$$\alpha = \cos^{-1}\left(\frac{\sin\theta}{\sin\theta_0}\cos\phi\right).$$

But we have seen that the incremental magnetic far field is given in terms of $\overline{A}$ by (6), so that expanding (6) in spherical coordinates

(22)
$$\overline{dH}_s(\overline{r})^{\,r\overset{\rightarrow\infty}{\sim}}dz'\left[A_x\sin\phi\hat\theta + (A_x\cos\theta\cos\phi - A_z\sin\theta)\hat\phi\right]$$

and hence

(23)
$$\overline{dH}_s(\overline{r})^{\,r\overset{\rightarrow\infty}{\sim}}dz'\frac{C}{C_0}\left\{\frac{H_{s\theta}(\phi\to\alpha)}{\sin\alpha}\sin\phi\hat\theta \right.$$
$$\left. + \left[\frac{H_{s\theta}(\phi\to\alpha)}{\sin\alpha}(\cos\phi\cos\theta + \cos\alpha\sin\theta\cot\theta_0) + H_{s\phi}(\phi\to\alpha)\frac{\sin\theta}{\sin\theta_0}\right]\hat\phi\right\}$$

The far electric field of the incremental current sheet is found from $\overline{dH}_s(\overline{r})$ through the spherical far-field relationship

(24)
$$\overline{dE}_s(\overline{r})^{\,r\overset{\rightarrow\infty}{\sim}} - Z_0\hat{r}\times\overline{dH}_s(\overline{r})$$

where Z_0 is the free-space impedance.

Expressions (23) and (24) for the incremental far field provide a substitution method for obtaining IDC's. Given a closed-form expression for the far field of a planar facet of a two-dimensional scatterer, we can find the IDC for the planar facet by direct substitution of the cylindrical far field components into a general expression. No knowledge of currents is needed and no integration is required.

The scattered fields of two-dimensional perfectly conducting cylinders (cylindrical in the general sense of having the same cross-section in planes perpendicular to the cylinder axis) divide conveniently into transverse electrical (TE) and transverse magnetic (TM) fields depending on whether the incident plane wave is TE ($E_z = 0$) or TM ($H_z = 0$), respectively. For the TE case the far field component $H_{s\phi}(E_{s\theta})$ is zero everywhere, while for the TM case $H_{s\theta}(E_{s\phi})$ is zero everywhere. Thus the incremental magnetic far field, (23), reduces respectively for these TE and TM cylindrical far fields to

(25)
$$\overline{dH}_s^{TE}(\overline{r})^{\,r\overset{\rightarrow\infty}{\sim}}dz'\frac{CH_{s\theta}(\phi\to\alpha)}{C_0\sin\alpha}\left[\sin\phi\hat\theta + (\cos\phi\cos\theta + \cos\alpha\sin\theta\cot\theta_0)\hat\phi\right]$$

and

(26)
$$\overline{dH}_s^{TM}(\overline{r})^{\,r\overset{\rightarrow\infty}{\sim}}dz'\frac{C}{C_0}H_{s\phi}(\phi\to\alpha)\frac{\sin\theta}{\sin\theta_0}\hat\phi.$$

The corresponding incremental electric far fields are given by

(27)
$$\overline{dE}_s^{TE}(\overline{r})^{\,r\overset{\rightarrow\infty}{\sim}}dz'\frac{CE_{s\phi}(\phi\to\alpha)}{C_0\sin\alpha}\left[\sin\phi\hat\phi - (\cos\phi\cos\theta + \cos\alpha\sin\theta\cot\theta_0)\hat\theta\right]$$

and

$$(28) \qquad \overline{dE}_s^{TM}(\overline{r}) \stackrel{r \to \infty}{\sim} dz' \frac{C}{C_0} E_{s\theta}(\phi \to \alpha) \frac{\sin \theta}{\sin \theta_0} \hat{\theta}$$

In these equations C and C_0 are given by (8) and (11) respectively, and α by (13).

As an example of the use of the general expressions we have just derived, we can obtain the IDC for a TM plane wave incident on a half-plane. The cylindrical far field of the half-plane thus illuminated may be found in textbooks or handbooks dealing with diffraction[5] (see Figure 4):

$$(29) \qquad \overline{E}_s \stackrel{\rho \to \infty}{\sim} -E_i \left(\frac{2}{\pi k \rho \sin \theta_0} \right)^{1/2} e^{i(k\rho \sin \theta_0 + \pi/4)} e^{-ikz \cos \theta_0} \frac{\sin \frac{\phi}{2} \sin \frac{\phi_0}{2}}{\cos \phi + \cos \phi_0} \hat{\theta}_0^\pi.$$

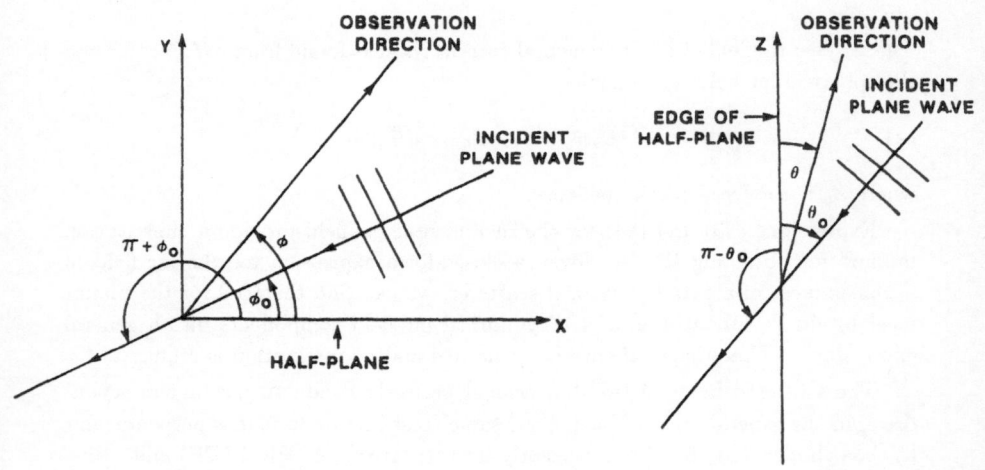

Figure 4. Geometry of a Half-Plane Illuminated by a Plane Wave.

If we substitute $E_{s\theta}$ into the general expression (28) we obtain for the halfplane TM IDC

$$(30) \qquad \overline{dE}_s^{TM}(\overline{r}) \stackrel{r \to \infty}{\sim} -dz' E_{iz} \frac{\sin \theta}{\sin^2 \theta_0} \frac{e^{ikr}}{4\pi r} \frac{4 \sin \frac{\alpha}{2} \sin \frac{\phi_0}{2}}{\cos \alpha + \cos \phi_0}.$$

This expression and expressions similarly obtained for the IDC's of the perfectly conducting infinite wedge for both TE and TM incident illumination agree with those obtained independently by Knott[3,6]. In addition to enabling IDC's for the perfectly conducting wedge to be obtained far more easily than hitherto, the substitution method for obtaining IDC's gave IDC's for the first time for the perfectly conducting strip and for the slit in an infinitesimally thin perfectly conducting plane.[4]

Before proceeding in the next section to look at some applications of IDC's, it should be mentioned that the important advantage that the integration of IDC's has over other techniques employed for taking into account scattering from edge discontinuities (e.g., GTD, PTD) is that the integration of IDC's yields corrections to the PO field that are valid in virtually all ranges of the pattern angles and thus avoids the necessity of employing a number of distinct formulations corresponding to different ranges.

IV. Some applications.

In this section we shall look briefly at two applications of IDC's, the first to the calculation of reflector antenna patterns, and the second to the calculation of scattering from a perfectly conducting thin disk illuminated by a plane wave.

Paraboloidal reflector antenna[7]. We consider a paraboloidal reflector antenna of focal length F and diameter D illuminated by a Huygens source located at the reflector focus. The rims of many reflectors are well approximated locally by perfectly conducting wedges. For simplicity here we consider the special, though important case of a reflector whose rim can be modeled locally by an infinitesimally thin, perfectly conducting half-plane. The correction to the PO reflector far field is thus obtained by integrating the nonuniform current IDC for the half-plane around the reflector rim. Computations were performed for a 20λ diameter paraboloidal reflector with $F/D = 0.4$. In the following plots, the primary antenna field is added to the secondary reflector field, and the main beam direction is $\theta = 180°$.

In Figure 5 we have plotted the forward region of the antenna pattern in the E-plane, with the dashed line showing the PO pattern and the solid line showing the pattern as corrected by adding the integral of the IDC. The principal effect of the inclusion of the nonuniform current at the reflector rim is a raising of the pattern nulls. The H-plane plots in the forward region, Figure 6, are quite similar. The more interesting plot, Figure 7, is that for the cross-polarized field in the plane making a 45° angle with the E- and H-plane. Here the strong influence of the nonuniform current is clearly seen. Comparison of our total (PO plus nonuniform plus feed) field patterns with the corresponding patterns obtained by Bach and Viskum[8] using a moment method, shows close agreement, in general throughout. As an example we have reproduced in Figure 8 the cross-polar amplitude pattern in the 45° plane obtained by Bach and Viskum[8]. (Note that in Bach and Viskum the main-beam direction is given by $\theta = 0°$.)

134

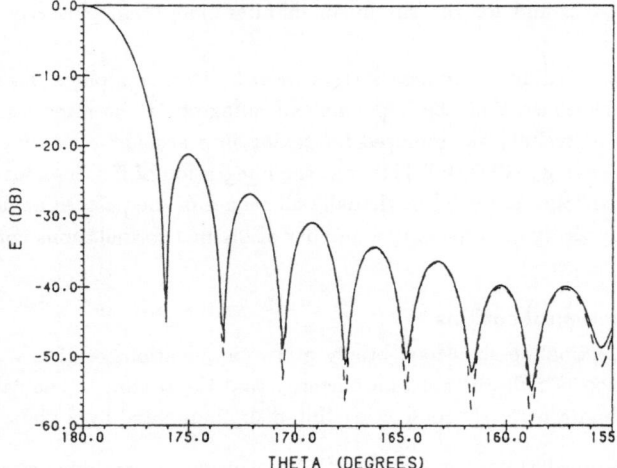

Figure 5. *E*-Plane Amplitude Pattern of 20λ Huygens-Feed Paraboloid Antenna with Primary Field Included; - - - - - PO, ——— PO + Nonuniform Current Field, $F/D = 0.4$, $\theta = [180°, 155°]$

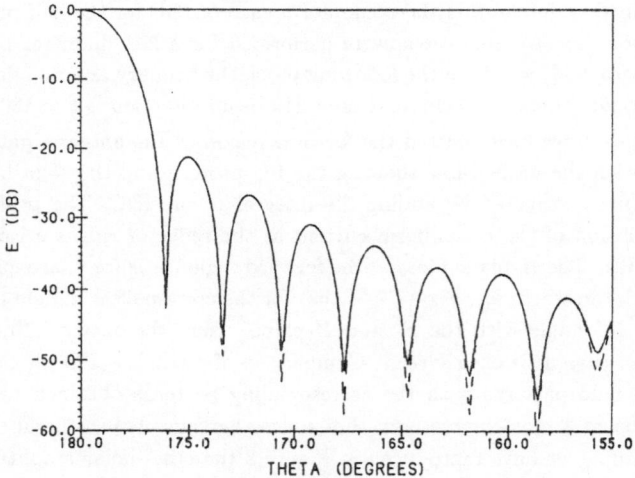

Figure 6. *H*-Plane Amplitude Pattern of 20λ Huygens-Feed Paraboloid Antenna with Primary Field Included; - - - - - PO, ——— PO + Nonuniform Current Field, $F/D = 0.4$, $\theta = [180°, 155°]$

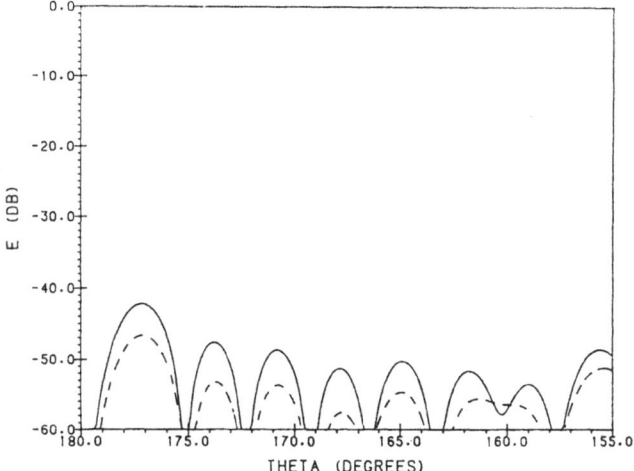

Figure 7. $\phi = 45°$ Cross-Polar Amplitude Pattern of 20λ Huygens-Feed Paraboloid Antenna with Primary Field Included; $- - - - -$ PO, ——— PO + Nonuniform Current Field, $F/D = 0.4, \theta = [180°, 155°]$

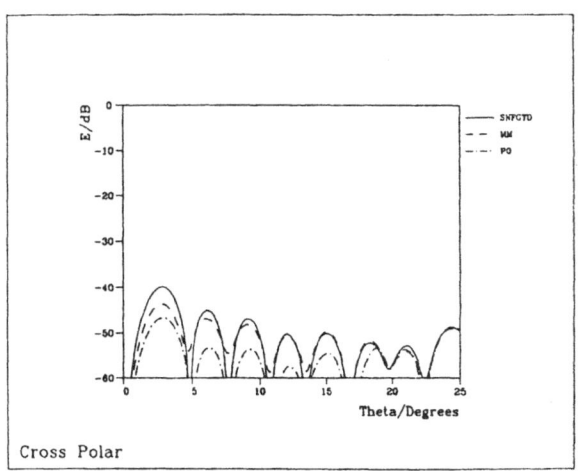

Figure 8. $\phi = 45°$ Cross-Polar Amplitude Pattern of 20λ Huygens-Feed Paraboloid Antenna with Primary Field Included; ——— Spherical Near-Field GTD, $- - - - -$ Moment Method, $- \cdot - \cdot - \cdot -$ PO. $F/D = 0.4$, (After Bach and Viskum[8])

Next, let us look at the patterns for the full range of θ, Figures 9-11. It is seen that the back lobes in the E-plane are considerably raised by including the nonuniform current field, while in the H-plane the back lobes are lowered by the nonuniform current field. The cross-polarized fields in the 45° plane are little affected in the back-lobe region by the nonuniform currents.

A further application of IDC's to reflector antenna pattern calculations is given in Reference 9 in which the IDC's of slits are used to calculate the effect of narrow cracks in the surface of reflector antennas, such as those resulting from the imperfect fitting together of panels to form large reflectors.

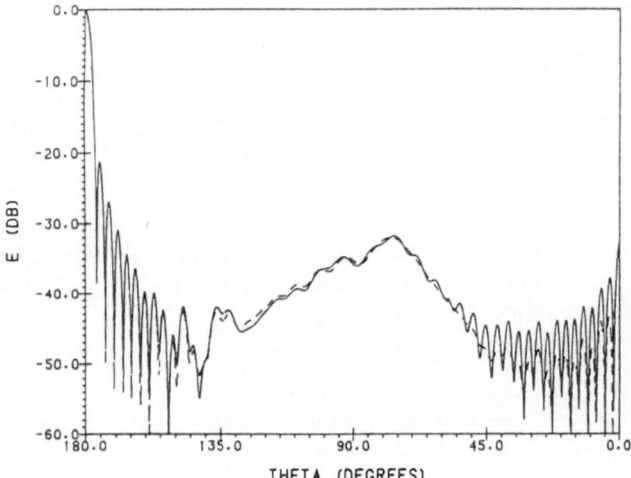

Figure 9. E-plane Amplitude Pattern of 20λ Huygens-Feed Paraboloid Antenna with Primary Field Included; – – – – – PO, —— PO + Nonuniform Current Field, $F/D = 0.4, \theta = [180°, 0°]$

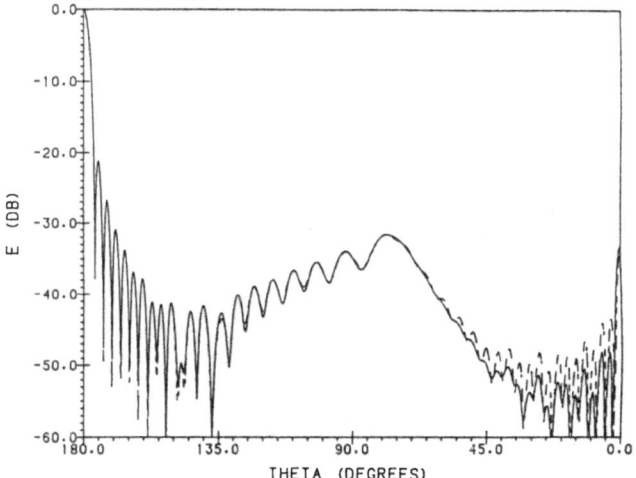

Figure 10. H-Plane Amplitude Pattern of 20λ Huygens-Feed Paraboloid Antenna with Primary Field Included; $- - - - -$ PO, ——— PO + Nonuniform Current Field, $F/D = 0.4, \theta = [180°, 0°]$

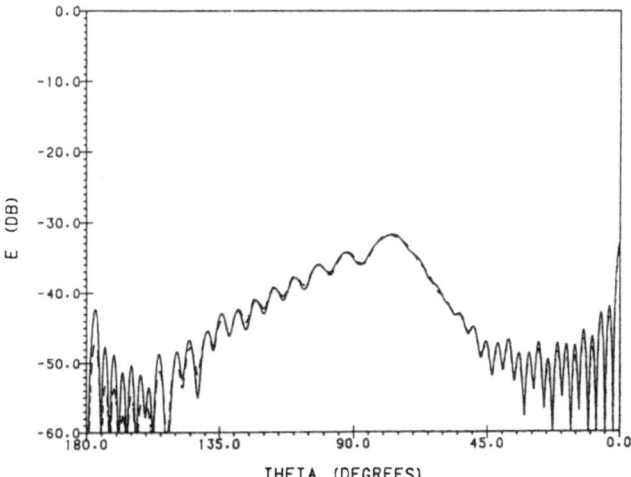

Figure 11. $\phi = 45°$ Cross-Polar Amplitude Pattern of 20λ Huygens-Feed Paraboloid Antenna with Primary Field Included; $- - - - -$ PO, ——— PO + Nonuniform Current Field, $F/D = 0.4, \theta = [180°, 0°]$

Scattering from a perfectly conducting disk[10]. We consider scattering from a disk of radius a, the geometry of which is shown in Figure 12. The primary source is a plane wave whose propagation direction lies in the xz-plane and makes an angle of θ_i with the z-axis. Two polarizations of the incident plane wave are considered: 1) **E** perpendicular to the plane of incidence (TE), and 2) **E** in the plane of incidence (TM). The far field scattered from the disk is obtained by adding to the PO far field the integral of the nonuniform current IDC of a strip (truncated half-plane) around the edge of the disk.

In Figure 13 we compare the PO solution and the exact solution (solid line) for backscattering from a circular disk of radius 2.4λ ($ka = 15$) with perpendicularly polarized plane-wave illumination. Note how poorly the PO solution approximates the exact solution as we move away from normal incidence to the disk. In contrast, in Figure 14, we compare the exact solution with the solution obtained by adding to the PO field the nonuniform current field approximated by integrating the IDC of a truncated half-plane around the disk circumference. Some small differences remain but the improvement is remarkable.

In Figure 15 we compare the exact and PO solutions for the back scatter pattern for parallel polarization of the incident plane wave. Again we note the increasing divergence of the two solutions as we move away from normal incidence. In contrast, in Figure 16 we compare the exact and corrected (IDC added to PO) solutions. The agreement is quite close throughout the range of incident angle.

The corresponding patterns for specular scattering are shown in Figures 17-20, again demonstrating the importance of the nonuniform current field in scattering calculations.

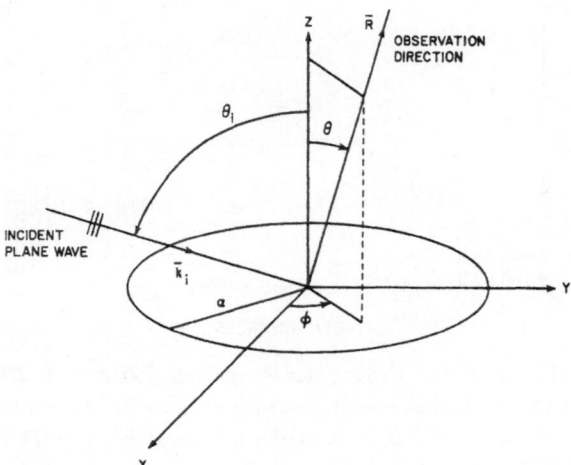

Figure 12. Geometry of Disk Illuminated by a Plane Wave

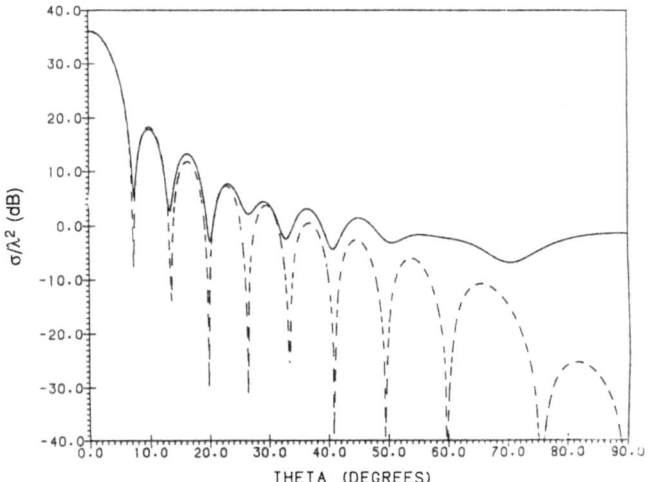

Figure 13. Back Scatter Cross Section Pattern of Disk, $ka = 15$, with Perpendicular Polarized Plane Wave Illumination; ——— Exact, $-----$ PO

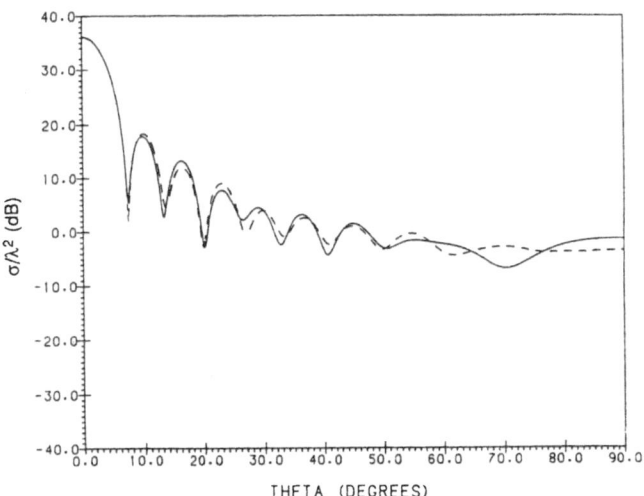

Figure 14. Back Scatter Cross Section Pattern of Disk, $ka = 15$, with Perpendicular Polarized Plane-Wave Illumination and Incremental Strips in Direction of Diffracted Rays; ——— Exact, $-----$ PO + Approximate Nonuniform Current Field Obtained by Integrating IDC

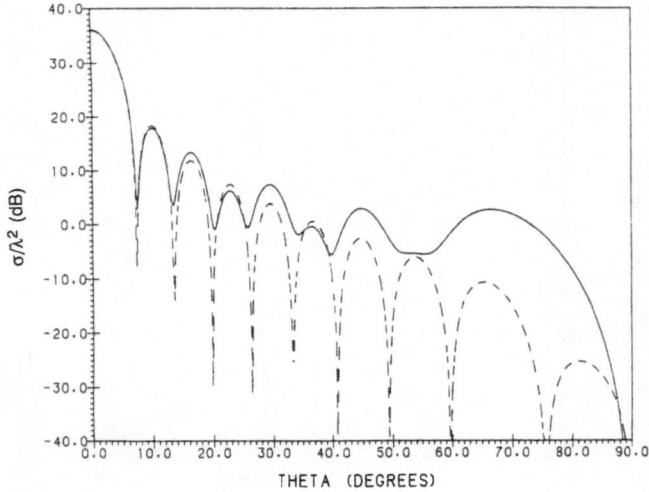

Figure 15. Back Scatter Cross Section Pattern of Disk, $ka = 15$, with Parallel Polarized Plane-Wave Illumination; ———— Exact, – – – – – PO

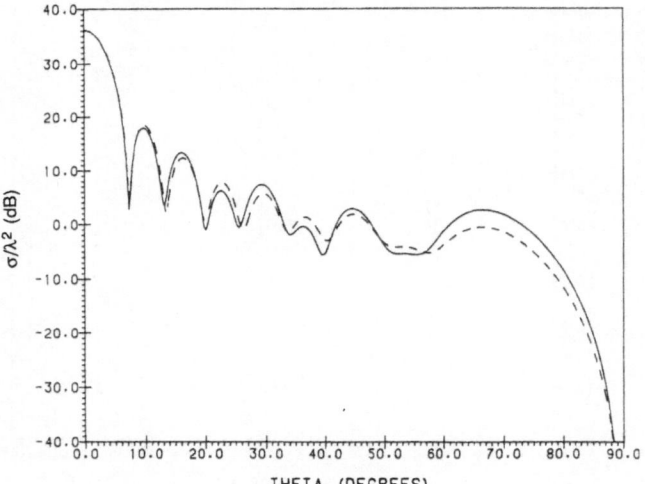

Figure 16. Back Scatter Cross Section Pattern of Disk, $ka = 15$, with Parallel Polarized Plane-Wave Illumination and Incremental Strips Parallel to x-Axis; ———— Exact, – – – – – PO + Approximate Nonuniform Current Field Obtained by Integrating IDC

141

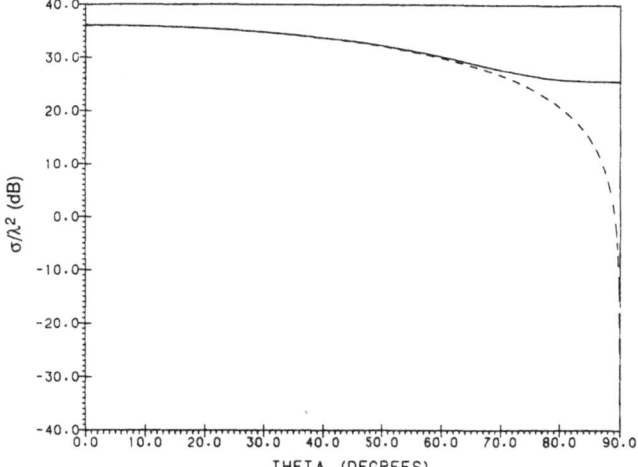

Figure 17. Specular Scatter Cross Section Pattern of Disk, $ka = 15$, with Perpendicular Polarized Plane-Wave Illumination; —— Exact, – – – – PO

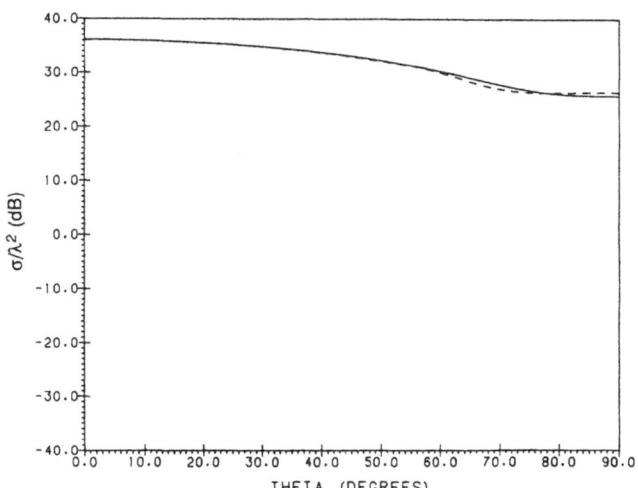

Figure 18. Specular Scatter Cross Section Pattern of Disk, $ka = 15$, with Perpendicular Polarized Plane-Wave Illumination and Incremental Strips in Direction of Diffracted Rays; —— Exact, – – – – PO

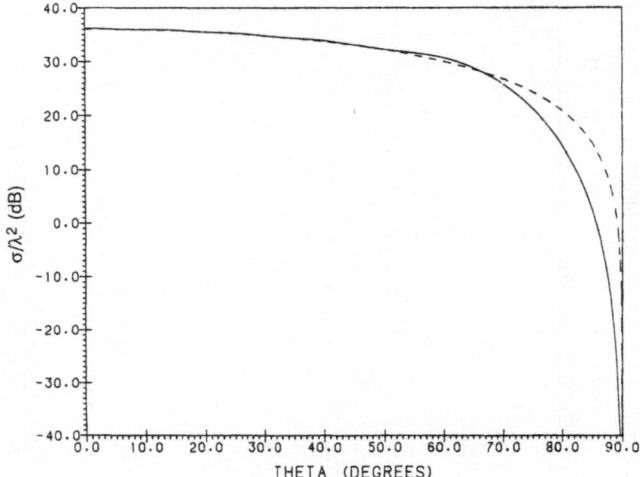

Figure 19. Specular Scatter Cross Section Pattern of Disk, $ka = 15$, with Parallel Polarized Plane-Wave Illumination; ——— Exact, – – – – – PO

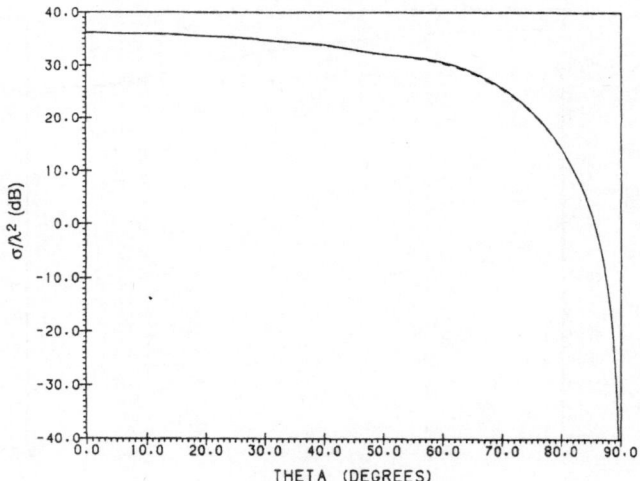

Figure 20. Specular Scatter Cross Section Pattern of Disk, $ka = 15$, with Parallel Polarized Plane-Wave Illumination and Incremental Strips Parallel to x-Axis; ——— Exact, – – – – – PO + Approximate Nonuniform Current Field Obtained by Integrating IDC

V. Some areas for further research.

In this final section we mention some problems still outstanding in obtaining and using IDC's.

So far in this paper we have tried to demonstrate that when IDC's can be obtained, they provide a highly useful technique for performing accurate scattering calculations. Unfortunately, at present IDC's are available for only planar, perfectly conducting canonical scatterers. It would be of much utility if a method not involving current integration could be found for obtaining IDC's for some curved canonical scatterers, such as the circular cylinder or parabolic cylinder (rounded wedge).[†] An extension to non-perfectly conducting scatterers would also be highly useful.

Still another problem involving IDC's arises in the procedure of using them. The IDC's are given in terms of a *local* coordinate system with origin at a point of an edge or shadow boundary of the scattering object. To integrate the IDC's, it is necessary to transform from the local coordinate system to the *global* coordinate system of the scatterer, incident illumination, and field point. The details of the transformation can be rather laborious. The attractiveness of using IDC's would be considerably enhanced if a general-purpose computer program could be written that circumvents the current necessity of working out the details of the coordinate transformation for each separate scattering configuration.

Finally, the method of integrating IDC's works well for electrically large scatterers when secondary diffraction is negligible. The determination of secondary diffraction from IDC's generally involves a prohibitively complicated double integration procedure.

REFERENCES

[1] K.M. MITZNER, Incremental Length Diffraction Coefficients, Technical Report No. AFAL-TR-73-296, Apr. 1974 (available from National Technical Information Service, Springfield, VA 22161, AD918861).

[2] A. MICHAELI, *Equivalent edge currents for arbitrary aspects of observation*, IEEE Trans. Antennas Propag., **AP-32**, pp. 252–258, Mar. 1984 (correction, **AP-33**, p. 227, Feb. 1985).

[3] E.F. KNOTT, *The relationship between Mitzner's ILDC and Michaeli's equivalent currents*, IEEE Trans. Antennas Propag., **AP-33**, pp. 112–114, Jan. 1985.

[4] R.A. SHORE AND A.D. YAGHJIAN, *Incremental diffraction coefficients for planar surfaces*, IEEE Trans. Antennas Propag., **AP-36**, pp. 55-70, Jan. 1988 (correction, **AP-37**, p. 1342, Oct. 1989); Also Incremental Diffraction Coefficients for Planar Surfaces, Part I: Theory, Technical Report RADC-TR-87-35, Apr. 1987 (available from National Technical Information Service, Springfield, VA 22161, ADA208595).

[5] J.J. BOWMAN AND T.B.A. SENIOR, *The half-plane*, Electromagnetic and Acoustic Scattering by Simple Shapes, J.J. Bowman, T.B.A. Senior, and P.L.E. Uslenghi, Eds. Amsterdam: North-Holland, 1969, ch. 4.

[6] E.F. KNOTT, J.F. SHAEFFER, AND M.T. TULEY, Radar Cross Section, Dedham, Mass: Artech House, 1985, ch. 5.

[7] R.A. SHORE AND A.D. YAGHJIAN, Incremental Diffraction Coefficients for Planar Surfaces, Part II: Calculation of the Nonuniform Current Correction to PO Reflector Antenna Patterns, Technical Report RADC-TR-87-213, Nov. 1987 (available from National Technical Information Services, Springfield, VA 222161, ADA208596).

[†]Recently, Hansen and Yaghjian[11] have made considerable progress in obtaining IDC's for arbitrary canonical scatterers of finite cross section.

[8] H. BACH AND H.H. VISKUM, *Comparison of three cross-polarization prediction methods for reflector antennas*, IEE Proc., **133**, pt. H, pp. 325-326, Aug. 1986. (See also *The SNFGTD method and its accuracy*, IEEE Trans. Antennas and Propag., **AP-35**, pp. 169-175, Feb. 1987.).

[9] R.A. SHORE AND A.D. YAGHJIAN, Incremental Diffraction Coefficients for Planar Surfaces, Part III: Pattern Effects of Narrow Cracks in the Surface of a Paraboloid Antenna, Technical Report RADC-TR-88-119, May 1988 (available from National Technical Information Services, Springfield, VA 22161, ADA207796).

[10] R.A. SHORE AND A.D. YAGHJIAN, Incremental Diffraction Coefficients for the Truncated Half-Plane and the Calculation of the Bistatic Radar Cross Section of the Disk, Technical Report RADC-TR-89-263, Oct. 1989 (available from National Technical Information Service, Springfield, VA 22161, ADA228689).

[11] T.B. HANSEN AND A.D. YAGHJIAN, Incremental Diffraction Coefficients for Cylinders of Arbitrary Cross Section: Applications to Diffraction from Ridges and Channels in Perfectly Conducting Surfaces, Digest of the IEEE AP-S Symposium, London, Ontario, Canada, June 1991.

A PROBLEM: HIGHER-ORDER SPECTRAL PHASE ESTIMATION

CHARLES L. WEIGEL*

Abstract. This paper briefly presents a problem of angle-of-arrival estimation using a bispectral technique and attempts to motivate the importance of the phase of a signal. Examining this problem leads to a more general question: How accurate is the phase of a spectral estimate? The purpose of this paper and the presentation on which it is based is to identify this more general problem in order to motivate for further analysis.

1.0 Introduction. The purpose of this paper is to motivate further study of the phase from higher-order spectral estimates. In the past, researchers have paid little attention to phase information, probably with good reason – power spectra do not have phase information. Higher-order spectra, however, do have significant phase information. For instance, one can use higher-order spectra to identify the magnitude and phase of a linear system. Also, phase information provides a window into the nonlinear characteristics of a signal which may lead to many other applications. Moreover, there are applications where only the phase is important: time-delay estimation, or angle-arrival estimation, holography, electron microscopy, among others. Because the phase of a signal does give useful information, understanding the accuracy of that phase estimate is important.

This paper presents angle-of-arrival estimation as an example of an application where the phase of a signal is significant. A bispectral technique is used to estimate the phase. In this case the accuracy of the angle-of-arrival estimate is related to the accuracy of the phase of the bispectral estimate. Understanding the accuracy of the phase estimate leads to an understanding of the accuracy of the angle-of-arrival estimate.

2.0 An Example: Angle-of-Arrival Estimation. The following figure shows the problem of angle-of-arrival estimation. A far-field source generates a uniform plane wave detected by a pair of sensors. The trigonometric equation shows the relation between the time delay of the signal and the arrival angle.

*Honeywell Inc., Underseas Systems Division, 600 Second Street N.E., Hopkins, MN 55343.

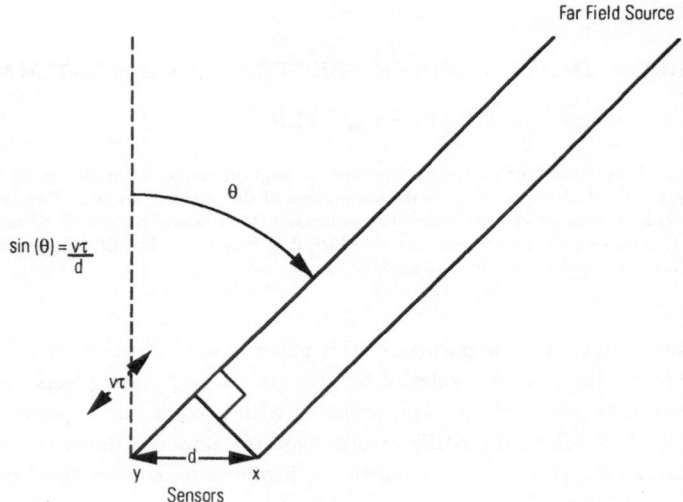

Adding noise to the signals brings another level of complication to the problem. The received signals are as follows:

(2.1)
$$x(t) = s(t) + n_x(t)$$

(2.2)
$$y(t) = s(t - \tau) + n_y(t)$$

where s is a stochastic noise source, s and (n_x, n_y) are statistically independent, $s, n_x,$ and n_y have zero means and n_x and n_y are correlated.

To estimate the angle-of-arrival θ, first estimate τ and then use the geometry to compute θ. The following equations outline a bispectral technique to estimate τ.

Calculate the bispectra:

(2.3)
$$B_{xxx}(\omega_1, \omega_2) = B_s(\omega_1, \omega_2) + N_{xxx}(\omega_1, \omega_2)$$

(2.4)
$$B_{xyx}(\omega_1, \omega_2) = B_s(\omega_1, \omega_2) \cdot e^{j\omega_1 \tau} + N_{xyx}(\omega_1, \omega_2)$$

Measure the phase difference:

(2.5)
$$\frac{B_{xyx}(\omega_1, \omega_2)}{B_{xxx}(\omega_1, \omega_2)} \approx e^{j\omega_1 \tau}$$

Average to find the time delay:

(2.6)
$$\hat{\tau} = N^{-1} \sum_{(\omega_1, \omega_2)} \omega^{-1} \arg \left[\frac{B_{xyx}(\omega_1, \omega_2)}{B_{xxx}(\omega_1, \omega_2)} \right]$$

Estimate the arrival angle:

$$(2.7) \qquad\qquad \theta = \sin^{-1}\left[\frac{\nu\tau}{d}\right]$$

The reason for looking at higher-order spectra is that if the corrupting noise is Gaussian, correlated or not, the bispectra of the noise are identically zero and the approximation in equation (2.5) becomes exact. Another way to solve this problem would be to delay and correlate x with y in the time domain. In this application, however, the time delay is always less than one sampling period and this time-domain correlation technique fails. Also, note that only the phase of the spectra appear in the estimate.

Because the estimate will not be perfect, one must ask how good is the arrival angle estimate (i.e. what are the statistics of $(\hat{\theta}-\theta)$) or how much data is required to make a good estimate. This question can be answered empirically, but an emperical answer does not give a good understanding of the estimate. Moreover it may not be feasible for some applications.

3.0 The Underlying Problem. Finding the statistics of the error on the angle-of-arrival estimate reduces to finding the statistics of the phase error of the estimate of the bispectrum. This problem relates not only to arrival angle estimation but to any use of the phase of the signal. Any bispectral estimation technique gives only an estimate $\hat{B}(\omega_1,\omega_2)$ which need not equal $B(\omega_1,\omega_2)$. Let

$$(3.1) \qquad\qquad \hat{B}(\omega_1,\omega_2) = |\hat{B}(\omega_1,\omega_2)| \cdot e^{j\cdot\mathrm{arg}(\hat{B}(\omega_1,\omega_2))}$$
$$(3.2) \qquad\qquad \Delta\phi = \mathrm{arg}(\hat{B}(\omega_1,\omega_2)) - \mathrm{arg}(B(\omega_1,\omega_2))$$

$\Delta\phi$ is the error of the phase of the bispectrum estimate and understanding this error is our goal.

The problem of estimating the bispectrum is not new, and some error analysis has been done. However, none of the analyses have addressed the phase of the estimate. For example, Van Ness published a paper demonstrating the asymptotic normality of bispectral estimates. In this work he divides the bispectrum into its real and imaginary parts and shows that each part converges to a normal distribution. [1] Van Ness comes close to solving the problem, but he does not consider a magnitude and phase representation. Likewise, Brillinger and Rosenblatt show the asymptotic convergence of all higher-order spectra but nothing about the intermediate statistics. [2] Also, T. Subba Rao shows that the estimate of the bispectrum converges on the actual bispectrum and gives a criteria for choosing a good windowing function for the estimator, but he only addresses the magnitude of the error and not its phase. [3]

4.0 Conclusion. The phase of signal does give useful information, not only in the arrival angle estimation example but also in other applications. With higher order spectra, the phase is meaningful and is almost surely an important key to understanding nonlinear characteristics revealed in higher-order spectra. In order to take advantage of phase information, however, its accuracy must be understood. The purpose of this paper is to highlight this need and to motivate research on the phase error of a spectral estimate. Some researchers have studied related problems, but no one has directly addressed this problem.

REFERENCES

[1] VAN NESS, *Asymptotic Normality of Estimates*, Ann. Math. Statist., vol. 37 (1966), pp. 1257–1272.
[2] D.R. BRILLINGER AND M. ROSENBLATT, *Asymptotic theory of estimates of k-th order spectra*, Spectral Analysis of Time Series, B. Harris, Ed. New York, NY: Wiley (1976), pp. 153–188.
[3] T. SUBBA RAO AND M.M. GABR, *An Introduction to Bispectal Analysis and Bilinear Time Series Models*, Springer-Verlag Lecture Notes in Statistics Series: vol. 24., (illus.) 280 p. 1984. pap. 17.00 (ISBN 0-387-96039-2).

DISPLACEMENT RANK AND DECONVOLUTION: APPLICATIONS AND OPEN PROBLEMS

HARPER J. WHITEHOUSE AND JEFFERY C. ALLEN[*]

Abstract. One method for obtaining a meaningful estimate of an input vector f from the noise-corrupted output vector $g = Af + \epsilon$ is by the method of regularization:

$$\hat{f} = Rg = (A^H A + \beta D^H D)^{-1} A^H g$$

where $\beta > 0$ and D are selected appropriate to the application. In this paper, it is assumed that the regularization matrix R is known but that many matrix-vector products Rg must be computed. This paper shows that if A and $D^H D$ have a small displacement ranks then a fast implementation of Rg is possible. In particular, if A and DD^H are $n \times n$ Toeplitz matrices, then Rg can be computed with $\mathcal{O}[4n \log n]$ multiplications.

1. Notation.

Definition 1.1. *The real numbers are denoted by* $\mathbf{R}$ *and the complex numbers by* $\mathbf{C}$. *The space of complex* $m \times n$ *matrices is denoted by* $\mathbf{C}^{m \times n}$ *and it is assumed throughout that* $m \geq n$.

1. *The conjugate transpose of a matrix* A *is denoted by* A^H.

2. I_n *denotes the* $n \times n$ *identity matrix*

3. *A matrix* $T \in \mathbf{C}^{m \times n}$ *is a Toeplitz matrix if it has the form*

$$T = \begin{bmatrix} t_0 & t_{-1} & t_{-2} & \cdots & t_{-n+1} \\ t_1 & t_0 & t_{-1} & \cdots & t_{-n+2} \\ t_2 & t_1 & t_0 & \cdots & t_{-n+3} \\ \vdots & \vdots & \vdots & \ddots & \vdots \\ t_{m-1} & t_{m-2} & t_{m-3} & \cdots & t_{m-n} \end{bmatrix}$$

4. *The lower* $m \times n$ *Toeplitz matrix* $L(x)$ *determined by the vector* $x \in \mathbf{C}^m$ *is given as*

$$L(x) = \begin{bmatrix} x_0 & 0 & 0 & \cdots & 0 & 0 \\ x_1 & x_0 & 0 & \cdots & 0 & 0 \\ x_2 & x_1 & x_0 & \cdots & 0 & 0 \\ \vdots & \vdots & \vdots & \ddots & x_0 & \vdots \\ x_{n-1} & x_{n-2} & x_{n-3} & \cdots & x_1 & x_0 \\ x_n & x_{n-1} & x_{n-2} & \cdots & x_2 & x_1 \\ \vdots & \vdots & \vdots & \ddots & x_0 & \vdots \\ x_{m-1} & x_{m-1} & x_{n-2} & \cdots & x_{m-n+1} & x_{m-n} \end{bmatrix}$$

An upper Toeplitz matrix $U(y)$ *is given as* $U(y) = L(y)^H$.

* Naval Ocean Systems Center San Diego, CA 92152-5000

5. Let T_{-m+1}, ..., T_{-1}, T_0, T_1, ..., T_{m-1} be $n \times n$ Toeplitz matrices. By a block Toeplitz matrix with Toeplitz blocks of type (m,n) is meant the $mn \times mn$ matrix

$$T = \begin{bmatrix} T_0 & T_{-1} & T_{-2} & \cdots & T_{-m+1} \\ T_1 & T_0 & T_{-1} & \cdots & T_{-m+2} \\ T_2 & T_1 & T_0 & \cdots & T_{-m+3} \\ \vdots & \vdots & \vdots & \ddots & \vdots \\ T_{m-1} & T_{m-2} & T_{m-3} & \cdots & T_0 \end{bmatrix}$$

The collection of all such matrices is denoted as $BTTB(m,n)$.

6. The $n \times n$ down-shift matrix Z_n is

$$Z_n = \begin{bmatrix} 0 & 0 & 0 & \cdots & 0 & 0 \\ 1 & 0 & 0 & \cdots & 0 & 0 \\ 0 & 1 & 0 & \cdots & 0 & 0 \\ \vdots & \vdots & \vdots & \cdots & \vdots & \vdots \\ 0 & 0 & 0 & \ddots & 0 & 0 \\ 0 & 0 & \cdots & \ddots & 1 & 0 \end{bmatrix}$$

7. For matrix $R \in \mathbf{C}^{n \times m}$, the $(+)$-displacement rank is the smallest integer $\alpha_+(R)$ for which

$$R = \sum_{k=1}^{\alpha_+(R)} L(x_k)U(y_k)$$

where $L(x_k) \in \mathbf{C}^{n \times n}$ is a lower Toeplitz matrix and $U(y_k) \in \mathbf{C}^{n \times m}$ is an upper Toeplitz matrix.

8. For matrix $R \in \mathbf{C}^{n \times m}$, the $(-)$-displacement rank is the smallest integer $\alpha_-(R)$ for which

$$R = \sum_{k=1}^{\alpha_-(R)} U(x_k)L(y_k)$$

where $U(x_k) \in \mathbf{C}^{n \times n}$ is an upper Toeplitz matrix and $L(y_k) \in \mathbf{C}^{n \times m}$ is a lower Toeplitz matrix.

9. The $(+)$-displacement operator Δ_+ is the linear map on $\mathbf{C}^{n \times m}$ given by

$$\Delta_+(R) = R - Z_n R Z_m^H$$

10. The $(-)$-displacement operator Δ_- is the linear map on $\mathbf{C}^{n \times m}$ given by

$$\Delta_-(R) = R - Z_n^H R Z_m$$

2. Introduction. One of the fundamental linear system models that occurs in many fields is given in equation (1). The system output g, the point-spread function

a, and the noise statistics of ϵ are known. The deconvolution problem is to recover the input f from the noisy observations g.

$$(1) \qquad g(x) = \int_{\mathbf{R}^d} a(x - y)f(y)\,dy + \epsilon(x)$$

where the dimension d is typically 1 or 2. The discrete version of equation (6) is the following vector equation

$$(2) \qquad g = Af + \epsilon$$

where $g \in \mathbf{C}^m$, $f \in \mathbf{C}^n$, and $\epsilon \in \mathbf{C}^m$. The matrix $A \in \mathbf{C}^{m \times n}$ is Toeplitz for $d = 1$ or block-Toeplitz with Toeplitz blocks for $d = 2$ [6]. Typically, deconvolution is accomplished by the *method of regularization* [9], [5]: to approximate f, find $f_{D,\beta}$ that minimizes the functional

$$\mathcal{L}(h) = ||g - Ah||^2 + \beta||Dh||^2$$

Here the matrix D and the regularization parameter $\beta > 0$ are chosen appropriate to the application. The regularized solution is given by

$$f_{D,\beta} = Rg = (A^H A + \beta D^H D)^{-1} A^H g$$

For the application motivating this paper [1], it is assumed that the regularization matrix R has been computed off-line but the experimenter is required to compute the matrix-vector multiplication Rg for many output vectors g of large length ($m \approx 1,000$). Therefore, efficient implementation of Rg is of considerable importance.

The basic idea for achieving a fast matrix-vector product when $d = 1$ is simple. Suppose the regularizing matrix R admits an expansion as the sum of products of lower and upper Toeplitz matrices:

$$(3) \qquad R = \sum_{k=0}^{\alpha_+(R)} L(x_k)U(y_k)$$

There are a number of algorithms that permit the fast computation of the products $L(x_k)U(y_k)g$ such as the Cooley-Tukey FFT, the Winograd prime factor DFT, the Fast Hartley transform, and number-theoretic transforms with $\mathcal{O}[m \log(m)]$ multiplications. From equation (3), we see that Rg may be calculated with $\mathcal{O}[\alpha_+(R)m \log(m)]$ multiplications. If both n and m are large (n, $m \approx 1,000$), and $\alpha_+(R)$ is small ($\alpha_+(R) \approx 4$) then the $\mathcal{O}[\alpha_+(R)m \log(m)]$ multiplications are substantially fewer than the nm multiplications usually required to compute Rg. Therefore, if the matrix R has a small displacement rank $\alpha_+(R)$ then a fast matrix-vector product may be obtained.

This paper marks our results to date on using the displacement rank to obtain an efficient means for calculating Rg. Section 3 shows that the regularization matrix R has displacement rank $\alpha_\pm(R) \leq 4$ when both A and D are Toeplitz matrices. Section 4 discusses the regularized deconvolution problem for $d = 2$, displays our current results and concludes with a collection of open problems that arise when considering two-dimensional deconvolution.

3. Displacement Rank for $d = 1$. This section undertakes a brief review of the relevant mathematical results concerning the displacement rank of a matrix to demonstrate the following generalization of our results obtained in [8]:

Theorem 3.1 *Let $A \in \mathbb{C}^{m \times n}$. Let $D \in \mathbb{C}^{n \times n}$ and strictly positive definite. Let $Q \in \mathbb{C}^{m \times m}$ be strictly positive definite. Then for all $\beta > 0$ there holds*

$$\alpha_{\pm}((A^H Q^{-1} A + \beta D)^{-1} A^H Q^{-1}) \leq \alpha_{\mp}(Q) + 2\alpha_{\mp}(A) + \alpha_{\mp}(D) + 2$$

If A, D and Q are Toeplitz, then

$$\alpha_{\pm}((A^H Q^{-1} A + \beta D)^{-1} A^H Q^{-1}) \leq 4$$

We note that the displacement operators $\Delta_{\pm}$ are closely related to the Stein transformation [4]. As operators on $\mathbb{C}^{n \times m}$, both are invertible as

$$\Delta_+^{-1}(X) = \sum_{k=0}^{n-1} Z_n^k X (Z_m^H)^k \text{ and } \Delta_-^{-1}(X) = \sum_{k=0}^{n-1} (Z_n^H)^k X Z_m^k$$

[7]. A key observation from these inversion formulas is that

$$L(x)U(y) = \Delta_+^{-1}(xy^H)$$

To apply this, let $\Delta_+(R)$ admit a singular-value decomposition (SVD) expansion of the form:

$$\Delta_+(R) = \sum_{k=1}^{\alpha} s_k u_k v_k^H$$

for singular values $s_1 \geq s_2 \geq \ldots \geq s_\alpha > 0$, orthonormal vectors $u_1, u_2 \ldots$, and $v_1, v_2 \ldots$. Applying Δ_+^{-1} to this SVD yields

$$R = \sum_{k=1}^{\alpha} s_k \Delta_+^{-1}(u_k v_k^H) = \sum_{k=1}^{\alpha} s_k L(u_k) U(v_k)$$

Therefore, $\alpha \geq \alpha_+(R)$. The invertibility of the displacement operators also implies the reverse inequality and we obtain the following alternative characterization of the displacement rank.

Lemma 3.1 [7] $\alpha_{\pm}(R) = \text{rank}(\Delta_{\pm}(R))$

From this characterization, it is straightforward to obtain the following collection of simple results.

Lemma 3.2 [7] *Let R and S be given matrices and let $\tilde{R}$ be any matrix which has R as a subblock. Then*

(a) $\alpha_{\pm}(R^H) = \alpha_{\pm}(R)$

(b) $\alpha_{\pm}(R) \leq \alpha_{\pm}(\tilde{R})$

(c) $\alpha_{\pm}(R + S) \leq \alpha_{\pm}(R) + \alpha_{\pm}(S)$

A fundamental result on displacement rank was obtained by Kailath, Kung and Morf in 1979.

Theorem 3.2 [7] *If matrix R is invertible then $\alpha_{\pm}(R) = \alpha_{\mp}(R^{-1})$*

We are now in a position to prove Theorem 3.1. Absorb β into matrix D. Start with the identity

$$\begin{bmatrix} -Q & A \\ A^H & D \end{bmatrix}^{-1} = \begin{bmatrix} -Q^{-1} + Q^{-1}A(D + A^H Q^{-1}A)^{-1}A^H Q^{-1} & Q^{-1}A(D + A^H Q^{-1}A)^{-1} \\ (D + A^H Q^{-1}A)^{-1}A^H Q^{-1} & (D + A^H Q^{-1}A)^{-1} \end{bmatrix}$$

Observe that $(D + A^H Q^{-1}A)^{-1}A^H Q^{-1}$ is the $(2,1)$ block of the preceding matrix. By Lemma 3.2, it follows that

$$\alpha_{\pm}((A^H Q^{-1}A + D)^{-1}A^H Q^{-1}) \leq \alpha_{\pm}\left(\begin{bmatrix} -Q & A \\ A^H & D \end{bmatrix}^{-1}\right)$$

By Theorem 3.2, it follows that

$$\alpha_{\pm}\left(\begin{bmatrix} -Q & A \\ A^H & D \end{bmatrix}^{-1}\right) = \alpha_{\mp}\left(\begin{bmatrix} -Q & A \\ A^H & D \end{bmatrix}\right)$$

By a straightforward calculation of the displacement rank of the preceding 2×2 block matrix, the reader can verify that the two inequalities given in Theorem 3.1 do indeed hold.

4. Displacement Rank for $d = 2$. When the discrete convolution of equation (2) is multivariate with $d = 2$, then matrix A is block Toeplitz with Toeplitz blocks [6]. If both the input and output vectors f and g arose from $m \times n$ digital images F and G as $f = \text{vec}(F)$ and $g = \text{vec}(G)$, then both have length mn, and $A \in \text{BTTB}(m,n)$. Exactly the same steps may be followed as in Section 3 for the proof of Theorem 3.1 to obtain the following result.

Theorem 4.1 *Suppose matrices A and D belong to $BTTB(m,n)$. If D is Hermitian and positive definite, then for all $\beta > 0$, $\alpha_{\pm}((A^H A + \beta D)^{-1}A^H) \leq 4m$*

Using the technique in [6], we can zero-pad the images F, G and the convolution function A to row size $N \geq m + n - 1$ and column size $N \geq m + n - 1$ and assume periodicity in both the rows and columns. Then the matrix-vector multiplication Af may be performed with a two-dimensional FFT by applying the one-dimensional FFT to the rows ($N \times N \log(N)$) and another one-dimensional FFT to the columns

$(N \times N \log(N))$ for a total of $\mathcal{O}[2N^2 \log(N)]$ multiplications. Therefore, Theorem 4.1 implies that regularization costs $\mathcal{O}[4N^3 \log(N)]$, assuming that $m \approx N/2$. Note that because $A \in \mathbf{C}^{mn \times mn}$, and assuming that $n \approx N/2$, a straight matrix-vector product costs $N^4/4$ multiplications.

5. Open Problems. The displacement rank of $4m$ for $A \in \mathrm{BTTB}(m,n)$ is not satisfying. We use this section to list two approaches that might reduce the cost of such a matrix-vector product. For the first approach, no mention has been made of the point-spread function a of equation (1) which determines the matrix A in equation (2) and how the form of the function could affect the decay of the singular values s_k of $\Delta_{\pm}(R)$. For example, suppose

$$\Delta_+(\mathcal{R}) = \sum_{k=1}^{\alpha(R)} s_k u_k v_k^H$$

is the SVD expansion. If an approximation of the form

$$R_\alpha = \sum_{k=1}^{\alpha} s_k L(u_k) U(v_k)$$

is used with $\alpha < \alpha_+(R)$, then Speiser [8] points out that the error incurred by regularizing with R_α instead of R is bounded by

$$\frac{s_{\alpha+1}}{2} \leq \|R_\alpha - R\| \leq n s_{\alpha+1}$$

If the singular values die off fast enough, these bounds indicate it might be sufficient to work with R_α rather than R. **How does the analytic structure of the point-spread function a affect the singular values of $\Delta_{\pm}(R)$?**

For the second approach, note that the displacement rank obtained in Theorem 4.1, while clearly generalizing the Theorem 3.1, did not really use the fact that matrix A was block Toeplitz with Toeplitz blocks. Kailath points out [7], [3] that the only requirement for a displacement-like representation is that the displacement matrix Z be nilpotent. For example, one might consider the Kronecker product

$$Z = Z_m \otimes I_n$$

to handle the $\mathrm{BTTB}(m,n)$ case. Examination of the definitions and results of Section 4 for this operator $Z_m \otimes I_n$ indicate that they readily generalize but do not lead to a reduction of the multiplication costs obtained in Section 4. **What displacement rank generalization will reduce the bounds obtained in Section 4?**

REFERENCES

[1] ABBISS, J. B., B. J. BRAMES, AND M. A. FIDDY, "Computational Aspects of a
     Regularized Image Reconstruction", to appear in *Digital Image Synthesis and Inverse
     Optics, Proceedings of SPIE, Volume 1351*, pp. 1351-09, San Diego, CA, July 9-13,
     1990.
[2] ABBISS, J. B., AND P. G. EARWICKER, "Compact Operator Equations, Regulariza-
     tion and Super-Resolution", *Proceedings of the IMA Conference on Mathematics and
     Signal Processing*, University of Bath, Bath, UK September 17-19, 1985.
[3] CHUN, J. AND T. KAILATH, *Displacement Representation and its Applications*, In-
     formation Systems Laboratory, Stanford University, Stanford, CA 94305
[4] DAVIS, C. W., *Elementary Divisors of the Stein Transformation*, SIAM Journal of
     Applied Mathematics, Volume 27, Number 3, November 1974, pp. 492-494
[5] GOLUB, G. H. AND C. F. VAN LOAN, *Matrix Computations, Second Edition*, The
     Johns Hopkins University Press, Baltimore and London, 1989
[6] GONZALEZ, R. C. AND PAUL WINTZ, *Digital Image Processing*, Addison-Wesley
     Publishing Company, 1977
[7] KAILATH, T., S.-Y. KUNG, AND M. MORF, *Displacement Ranks of Matrices and
     Linear Equations*, Journal of Mathematical Analysis and Applications, Volume 68,
     pages 395-407, 1979
[8] SPEISER, J. M., H. J. WHITEHOUSE, AND J. C. ALLEN, "Fast Matrix-Vector Mul-
     tiplication Using Displacement Rank Approximation via an SVD", to appear in the
     Proceedings of the Second International Workshop on SVD and Signal Processing
     held at the University of Rhode Island, June 25-27, 1990
[9] WAHBA, G., *Practical Approximate Solutions to Linear Operator Equations When
     the Data are Noisy*, SIAM Journal of Numerical Analysis, Volume 14, Number 4,
     September 1977, pages 651-667.